T0174854

Corrosion Control
Through Organic Coatings

CORROSION TECHNOLOGY

Editor
Philip A. Schweitzer, P.E.
Consultant
York, Pennsylvania

Corrosion Control Through Organic Coatings, Second Edition

By
Ole Øystein Knudsen and Amy Forsgren

CRC Press
Taylor & Francis Group
Boca Raton London New York

CRC Press is an imprint of the
Taylor & Francis Group, an **informa** business

CRC Press
Taylor & Francis Group
6000 Broken Sound Parkway NW, Suite 300
Boca Raton, FL 33487-2742

First issued in paperback 2019

© 2017 by Taylor & Francis Group, LLC
CRC Press is an imprint of Taylor & Francis Group, an Informa business

No claim to original U.S. Government works

ISBN-13: 978-1-4987-6072-0 (hbk)
ISBN-13: 978-0-367-87711-8 (pbk)

This book contains information obtained from authentic and highly regarded sources. Reasonable efforts have been made to publish reliable data and information, but the author and publisher cannot assume responsibility for the validity of all materials or the consequences of their use. The authors and publishers have attempted to trace the copyright holders of all material reproduced in this publication and apologize to copyright holders if permission to publish in this form has not been obtained. If any copyright material has not been acknowledged please write and let us know so we may rectify in any future reprint.

Except as permitted under U.S. Copyright Law, no part of this book may be reprinted, reproduced, transmitted, or utilized in any form by any electronic, mechanical, or other means, now known or hereafter invented, including photocopying, microfilming, and recording, or in any information storage or retrieval system, without written permission from the publishers.

For permission to photocopy or use material electronically from this work, please access www.copyright.com (http://www.copyright.com/) or contact the Copyright Clearance Center, Inc. (CCC), 222 Rosewood Drive, Danvers, MA 01923, 978-750-8400. CCC is a not-for-profit organization that provides licenses and registration for a variety of users. For organizations that have been granted a photocopy license by the CCC, a separate system of payment has been arranged.

Trademark Notice: Product or corporate names may be trademarks or registered trademarks, and are used only for identification and explanation without intent to infringe.

Library of Congress Cataloging-in-Publication Data

Names: Forsgren, Amy, author. | Knudsen, Ole Øystein, author.
Title: Corrosion control through organic coatings / Ole Øystein Knudsen, Amy Forsgren.
Description: Second edition. | Boca Raton : a CRC title, part of the Taylor & Francis imprint, a member of the Taylor & Francis Group, the Academic Division of T&F Informa, plc, [2017] | Series: Corrosion technology | Revised edition of: Corrosion control through organic coatings / Amy Forsgren. 2006 | Includes index.
Identifiers: LCCN 2016040263| ISBN 9781498760720 (hardback : alk. paper) | ISBN 9781315153186 (ebook)
Subjects: LCSH: Protective coatings. | Corrosion and anti-corrosives. | Organic compounds.
Classification: LCC TA418.76 .F67 2017 | DDC 620.1/1223--dc23
LC record available at https://lccn.loc.gov/2016040263

Visit the Taylor & Francis Web site at
http://www.taylorandfrancis.com

and the CRC Press Web site at
http://www.crcpress.com

Dedication

To Tuva, Njål, and Eirin—thanks for your support and encouragement.

Contents

Preface

Since Amy Forsgren wrote the first edition of this book, her career has taken another direction and she is now working with water and wastewater transport and treatment technology. When Taylor & Francis contacted her for preparing a revision, she asked me to take responsibility for the book, which I accepted, evidently. She has, however, still been involved in the preparation of this edition.

I have expanded the book with new chapters on coating degradation, protective properties, duplex coatings powder coatings, and chemical pretreatment, which are subjects on which I have worked. Other subjects, such as surface preparation standards, coating selection, and application, methods are still not covered, but information is readily available elsewhere. The contents of the first edition have been restructured and updated, but all the subjects treated in the first edition are still included. Chapter 8 on heavy metal contamination of blasting media is, for example, still depressingly relevant.

Our ambition has been to write a book that covers the practical aspects of corrosion protection with organic coatings, and link this to the ongoing research and development. This is a challenging task, due to the continuous development of new coating technology and the vast amount of research published on this topic every year. However, this is also why this book is relevant and worth revising.

We have tried to write this book for everybody working with protective organic coatings, for example:

- Maintenance engineers who specify or use anticorrosion paints and need a sound working knowledge of different coating types and some orientation in how to test coatings for corrosion protection
- Buyers or specifiers of coatings, who need to know quickly which tests provide useful knowledge about performance and which do not
- Researchers working with coating degradation and accelerated test methods, who need an in-depth knowledge of aging mechanisms of coatings, in order to develop more accurate tests
- Applicators interested in providing safe working environments for personnel performing surface preparation
- Owners of older steel structures who find themselves faced with removal of lead-based paint when carrying out maintenance painting

As a professor at the Norwegian University of Science and Technology, teaching coatings technology, I have wished for a textbook on protective organic coatings for students at the BSc and MSc levels. The preparation of the second edition of this book has also been made with this in mind.

Ole Øystein Knudsen
Trondheim

Acknowledgments

Without the help of many people, this book would not have been possible. We wish in particular to thank Lars Krantz, Tone Heggenougen, and Njål Knudsen for generously creating the illustrations. Thanks to Håvard Undrum at Jotun, Lars Erik Owe at Jotun Powder Coatings, and Otto Lunder at SINTEF for comments to Chapters 3, 6, and 9, respectively.

Authors

Ole Øystein Knudsen received his MSc in chemical engineering at the Norwegian Institute of Technology in 1990, and his PhD in 1998 from the Norwegian University of Science and Technology with a thesis on cathodic disbonding. Since 1998, he has been working with research on coating degradation and corrosion at SINTEF. Since 2008, he has been an adjunct professor at the Norwegian University of Science and Technology, teaching about protective coatings. Dr. Knudsen lives in Trondheim with his family.

Amy Forsgren received her chemical engineering education at the University of Cincinnati in Ohio in 1986. She then did research in coatings for the paper industry for 3 years, before moving to Detroit, Michigan. There, she spent 6 years in anticorrosion coatings research at Ford Motor Company, before returning to Sweden in 1996 to lead the protective coatings program at the Swedish Corrosion Institute. She is now working in the water and wastewater industry at Xylem Inc. in Stockholm. Mrs. Forsgren lives in Stockholm with her family.

1 Introduction

This book is not about corrosion; rather, it is about paints that prevent corrosion. It was written for those who must protect structural steel from rusting by using anticorrosion paints.

Selecting an adequately protective coating for a construction can be a difficult task, as has been demonstrated by many expensive mistakes. Some of them are included in this book, since they give us valuable insights about degradation mechanisms and potential threats to our coating. Repairing coatings in the field is usually a lot more expensive than in the yard due to access problems. In addition, the repair coating rarely has the same lifetime as the original coating. Thus, selecting the right coating and applying it correctly the first time is crucial for the life cycle cost of corrosion protection.

A coating is rarely applied for a single purpose only. Decorative properties are important for almost all constructions. An aesthetic appearance signals care and maintenance, and gives us a feeling of safety, in addition to a better visual experience. Other functionalities a paint coating can deliver are signal colors, thermal insulation, friction, nonstick properties, electric insulation, and so forth. Such requirements will put additional restrictions on the coating selection.

1.1 SCOPE OF THE BOOK

The scope of this book is heavy-duty protective coatings used to protect structural steel, infrastructure components made of steel, and heavy steel process equipment. The areas covered by this book have been chosen to reflect the daily concerns and choices faced by engineers who use heavy-duty coating, including

- Composition of anticorrosion coatings
- Waterborne and powder coatings
- Blast cleaning and other heavy surface pretreatments
- Abrasive blasting and heavy metal contamination
- Weathering and aging of paint
- Degradation of paint by corrosion reactions
- Corrosion testing—background and theoretical considerations
- Corrosion testing—practice

You do not need detailed knowledge about corrosion to read this book, but a certain knowledge about the basics of corrosion will be helpful. Corrosion theory will not be explained here; instead, reference is made to other excellent books on corrosion [1–3].

1.2 TARGET GROUP DESCRIPTION

The target group for this book consists of those who specify, formulate, test, or do research in heavy-duty coatings for such applications as

- Marine vessels and offshore installations
- Onshore and offshore pipelines
- Boxes and girders used under bridges or metal gratings used in the decks of bridges
- Penstock pipes in hydropower plants
- Tanks for chemical storage, potable water, or waste treatment
- Handrails for concrete steps in the fronts of buildings
- Masts for telecommunications antennas
- Power line pylons
- Beams in the roof and walls of food processing plants
- Grating and framework around processing equipment in paper mills

All these forms of structural steel have at least two things in common:

1. Given a chance, the iron in them will turn to iron oxide.
2. When the steel begins rusting, it cannot be pulled out of service and sent back to a factory for treatment.

During the service life of one of these structures, maintenance painting will have to be done on-site. This imposes certain limitations on the choices the maintenance engineer can make. Coatings that must be applied in a factory cannot be reapplied once the steel is in service. This precludes certain organic paints, such as powder coatings or electrodeposition coatings, and several inorganic pretreatments, such as phosphating, hot-dip galvanizing, and chromating. New construction can commonly be protected with these coatings, but they are almost always a one-time-only treatment. When the steel has been in service for a number of years and maintenance coating is being considered, the number of practical techniques is narrowed. This is not to say that the maintenance engineer must face corrosion empty-handed; more good paints are available now than ever before, and the number of feasible pretreatments for cleaning steel *in situ* is growing. In addition, coatings users now face such pressures as environmental responsibility in choosing new coatings and disposing of spent abrasives, as well as increased awareness of health hazards associated with certain pretreatment methods.

1.3 COATED METAL SYSTEM

Steel is our most important construction metal, and most of this book is devoted to protective coatings for steel. However, other metals are also painted.

Most aluminum alloys used as construction metals are much less susceptible to corrosion than steel, and paint coatings are in most cases applied for aesthetical reasons and not corrosion protection. The coatings applied on aluminum are therefore usually thinner than the coatings applied on steel. Since aluminum has different corrosion properties than steel, most of the pretreatment processes applied are also different. Coatings on aluminum also differ from steel with respect to degradation mechanisms.

The third metal that is important with respect to organic coatings is zinc, that is, zinc-coated steel. The steel is first hot-dip galvanized, electrogalvanized, or thermally sprayed with zinc, typically to get 10–100 μm of metallic zinc on the surface. The zinc coating is then coated with an organic coating. Such coating systems, called duplex coatings, give excellent corrosion protection, and may have extremely long life with a minimum of maintenance. For constructions in corrosive environments that will be used for several decades, for example, road bridges and offshore installations, such coating systems may give lower life cycle costs than traditional paint systems. The latter usually give lower costs in the construction phase, but when the maintenance costs expected during the life of the structure are included in the calculation, the duplex coating system is often favorable.

The organic polymeric coating system may vary in complexity from the simple barrier coating on a blast-cleaned surface to a combination of conversion coating, primer, intermediate coats, and topcoat.

A coating system consisting of three different coats is illustrated in Figure 1.1. The coats are

1. A primer with metallic zinc particles. The function of the zinc is to provide cathodic protection. Being the first coat, the primer also provides adhesion to the substrate.

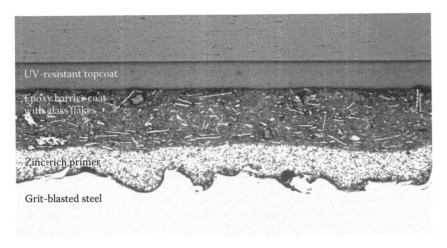

FIGURE 1.1 Blast-cleaned steel with a three-coat paint system—zinc-rich primer, barrier coat with flake-shaped pigments, and UV-resistant topcoat. The cross section image was taken with an electron microscope. A range of different qualities may be obtained by varying the pretreatment of the metal surface and the coating system (generic type, thickness, and number of coats).

2. The intermediate coat is typically a barrier coating whose main action is to limit the transport of ions to the metal surface. The transport of oxygen and water is also limited.
3. The topcoat is, if necessary, formulated to protect against degradation by ultraviolet (UV) light. In addition, the topcoat will give the required gloss, opacity, and color, that is, the visual appearance. It is also an additional barrier.

Hence, the individual coats have different purposes and vary in formulation in order to give different properties to the coating system. By choosing different types of coats, thicknesses and number of coats, the properties and expected lifetime of the coating system is changed.

REFERENCES

1. Revie, R.W., and H.H. Uhlig. *Corrosion and Corrosion Control: An Introduction to Corrosion Science and Engineering*. 4th ed. Hoboken, NJ: John Wiley & Sons, 2008.
2. McCafferty, E. *Introduction to Corrosion Science*. New York: Springer-Verlag, 2010.
3. Bardal, E. *Corrosion and Protection*. London: Springer, 2003.

2 Protection Mechanisms of Organic Coatings

In this chapter, we see that there are primarily three mechanisms by which organic coatings protect against corrosion:

1. Stabilizing the passivating surface oxide
2. Cathodic protection
3. Passivating pigments

The barrier and adhesion properties of the coating are not protection mechanisms in themselves, perhaps a bit surprisingly, but that does not mean that these properties are unimportant. On the contrary, they are vital for the longevity of the coating. Poor adhesion or barrier properties results in rapid failure of the coating and loss of protection.

2.1 BARRIER AGAINST OXYGEN AND WATER

Intuitively, we think that organic coatings protect the substrate from corrosion by being barriers against oxygen and water. Organic coatings are barriers against oxygen and water, but they are not impermeable to these species. Table 2.1 shows the permeation rates of water vapor and oxygen through several coatings as measured by Thomas [1,2]. Except for the aluminum epoxy mastic, the coatings in the tables are outdated and not in use anymore. However, they are all protective organic barrier coatings. A simple calculation will show that in order to maintain a corrosion rate of 100 μm/year on steel, we will need a water permeation of 0.93 g/m^2/day and an oxygen permeation of 575 cm^3/m^2/day [1,2]. The measured water and oxygen permeabilities given in Table 2.1 should result in a considerable corrosion rate under all the listed coatings, but that is not what was observed. All three coatings provided excellent corrosion protection. The barrier explanation for corrosion protection cannot be correct.

2.2 STABILIZING THE PASSIVATING SURFACE OXIDE

Barrier properties against water and oxygen cannot explain the protective effect of the coating. Instead, we must look to the corrosion theory to see how corrosion can stop when all the necessary reactants actually are present. The phenomenon is called passivity; that is, protective surface oxides stop the electrochemical reaction between oxygen and the metal. A short overview of passivity is given in this section. The classic theory on passivity by Cabrera and Mott [3] or the point defect model by Macdonald [4] provides the details.

TABLE 2.1

Water Vapor and Oxygen Permeability

Coating Type	Water Vapor Permeability, $g/m^2/25 \ \mu m/day$	Oxygen Permeability, $cc/m^2/100 \ \mu m/day$
Chlorinated rubber	20 ± 3	30 ± 7
Coat tar epoxy	30 ± 1	213 ± 38
Aluminum epoxy mastic	42 ± 6	110 ± 37

Source: Thomas, N.L., *Prog. Org. Coat.*, 19, 101, 1991; Thomas, N.L., in *Proceedings of the Symposium on Advances in Corrosion Protection by Organic Coatings*, Electrochemical Society Pennington, NJ, 1989, 451.

When paint is applied on a clean metal surface, the paint is actually not applied on the metal, but on a tight, adhering oxide on the metal surface. The oxide is immediately formed when a new metal surface is created, for example, during blast cleaning, in reactions between the metal and oxygen and humidity in the air. This oxide separates the metal from the environment.

Corrosion is the combination of an anodic reaction and a cathodic reaction. The anodic reaction is the oxidation of the metal, for example, iron:

$$Fe = Fe^{2+} + 2e^- \tag{2.1}$$

The anodic reaction must take place under the oxide at the oxide–metal interface, where the metal atoms are. The cathodic reaction is oxygen reduction:

$$O_2 + 2H_2O + 4e^- = 4OH^- \tag{2.2}$$

or hydrogen evolution:

$$2H_2O + 2e^- = H_2 + 2OH^- \tag{2.3}$$

The cathodic reactions will take place on top of the oxide, where oxygen and water meet the material surface. Hence, the anodic and cathodic reactions are separated by the oxide, as illustrated in Figure 2.1. In order for the reactions to occur, two premises must be met:

1. The electrons that are liberated in the anodic reaction under the oxide must travel to the surface of the oxide, where they are consumed in the cathodic reaction. Some oxides are electron conductors (e.g., Fe_3O_4), some are semiconductors (e.g., Fe_2O_3) and some are insulators. The electrons can also be transported through the oxide by tunneling [3].
2. Ions must also be transported through the oxide in order to maintain charge neutrality. Whether it is the metal ions, the oxygen ions, or both that are transported depends on the metal and type of oxide.

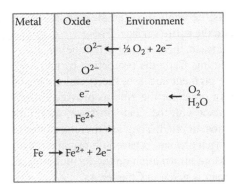

FIGURE 2.1 Schematic illustration of the anodic and cathodic reactions at the metal–oxide and oxide–coating interfaces, and transport of electrons and ions.

As the reactions continue, the oxide grows in thickness. However, as the oxide becomes thicker, the distance between the anodic reaction under the oxide and the cathodic reaction on top of the oxide increases. This increases the resistance against the transport of electrons and ions through the oxide, and the reaction rate will gradually decrease. The oxide thickness will follow a logarithmic growth rate [3]. The reaction rate will decrease and the metal is passivated.

In an aggressive environment, the oxide is continuously dissolved from the surface and the metal will corrode, for example, when steel is exposed in water. The dissolution and transformation of the oxide prevent it from becoming protective. Stainless steel and aluminum are covered by chromium oxide and aluminum oxide, respectively. These oxides are insoluble and stable, which is why these metals are corrosion resistant. Actually, these metals will also corrode in any environment where the oxide is destabilized, for example, when aluminum is exposed to acidic or alkaline solutions. Iron, however, does not form fully protective oxides in wet environments.

On a painted steel surface, there will be no water phase to dissolve the surface oxide. Any reactions caused by water and oxygen penetrating the paint film will contribute to oxide growth, until the oxide is too thick for the reactions to continue. The steel is passivated, and will remain so, because the paint film preserves the oxide. Hence, the most important protection mechanism for organic coatings is that the coating protects the surface oxide, which actually is responsible for the corrosion protection.

The mechanism has not been proven by experimental evidence yet, but it is in agreement with previous investigations and corrosion theory in general. A similar mechanism has also been suggested by Mills and Jamali [5].

2.3 CATHODIC PROTECTION

Steel that is immersed in an electrolyte may be protected by sacrificial anodes mounted on the structure. This will not work in atmosphere, since there is no electrolytic contact between the anode and the steel. To overcome this limitation, the sacrificial material must be applied to the steel as a coating, which is in electric contact with the steel on the entire surface. There are many ways of applying a sacrificial metal on a steel surface, the most important being hot-dip galvanizing, thermal spraying, and electroplating. Cathodic protection by painting is achieved with zinc-rich paints, whose zinc pigment acts as a sacrificial anode, corroding preferentially to the steel substrate. In order for the zinc to provide cathodic protection, the zinc must be in electric contact with the steel substrate, which means that the zinc-rich paint must be the first coat applied. They are therefore referred to as zinc-rich primers. The binder in zinc-rich primers is based on organic polymers, typically epoxies, or inorganic silicates. More information on the formulation of zinc-rich coatings and the protection mechanism is given in Chapter 4.

2.4   PASSIVATING THE SUBSTRATE WITH PIGMENTS

The passivation mechanism described above can be enhanced by pigmenting the paint with reactants or species that precipitate on the metal surface, strengthening or adding to the protection of the original oxide. Anticorrosion pigments in a coating are slightly soluble in water. Their dissociated ions migrate to the coating–metal interface and passivate it by supporting the formation of thin layers of insoluble corrosion products, which inhibit further corrosion [6–8]. Historically, hexavalent chromium (chromate) has often been used for this purpose, since chromate is very effective and gives highly protective coatings. However, chromates are also highly toxic and carcinogenic, and are now only used in special cases (in Europe and North America). They have been banned in certain industries, for example, the European automotive industry through a European Union directive [9]. Today, the most important passivating pigments are phosphate salts [10]. These are not considered toxic, but they are also less effective than the chromates.

For more information about anticorrosion pigments, see Chapter 4.

2.5   DURABLE PROTECTION

As stated in the beginning of this chapter, barrier and adhesion should not be considered protection mechanisms, although they are key properties of the coating. Poor barrier or adhesion properties may result in very rapid degradation of the coating, followed by corrosion of the substrate. The importance of barrier and adhesion for coating durability is briefly discussed below, while details about these properties are discussed in more detail in Chapter 10.

As shown in Table 2.1, coatings are permeable to oxygen and water, but may still provide excellent corrosion protection. However, the coating must be a good barrier against ions. When ions penetrate the coating, the passivation mechanism described in Section 2.2 will fail. If cations migrate through the coating from the outside, the cathodic reaction will start and hydroxides will be formed. The ions may cause osmosis and blistering of the coating. Salt diffusing through the coating will have the same effect. This is why coatings that work by stabilizing the passivating oxide often are referred to as barrier coatings. In effect, protection is lost. This is discussed further in Chapter 12. This is supported by the results of many investigations, showing that ion penetration of the coating is the first step in the degradation of a seemingly intact organic coating [11–13]. The correlation between ionic penetration and degradation means that coating resistance can be used as an indicator for coating performance. Both alternating current (AC) and direct current (DC) methods can be used for measuring coating resistance [14,15]. Protective barrier coatings on steel normally have a resistance above 10^6 Ωcm^2 [5,16].

The role of adhesion is to create the necessary conditions so that corrosion protection mechanisms can work. A coating cannot stabilize the protective oxide or prevent ions from reaching the metal surface, unless it is in intimate contact— at the atomic level— with the surface. The more chemical bonds between the surface and coating, the closer the contact and the stronger the adhesion. An irreverent view could be that the higher the number of sites on the metal that are taken up in bonding with the coating, the lower the number of sites remaining available for electrochemical mischief, or as Koehler expressed it,

> The position taken here is that from a corrosion standpoint, the degree of adhesion is in itself not important. It is only important that some degree of adhesion to the metal substrate be maintained. Naturally, if some external agency causes detachment of the organic coating and there is a concurrent break in the organic coating, the coating will no longer serve its function over the affected area. Typically, however, the detachment occurring is the result of the corrosion processes and is not quantitatively related to adhesion. [17]

In summary, good adhesion of the coating to the substrate could be described as a necessary but not sufficient condition for good corrosion protection. For all the protection mechanisms described in the previous sections, good adhesion of the coating to the metal is a necessary condition. However, good adhesion alone is not enough; adhesion tests in isolation cannot predict the ability of a coating to control corrosion [8].

REFERENCES

1. Thomas, N.L. *Prog. Org. Coat.* 19, 101, 1991.
2. Thomas, N.L. The protective action of red lead pigmented alkyds on rusted mild steel, in *Proceedings of the Symposium on Advances in Corrosion Protection by Organic Coatings*, Electrochemical Society, Pennington, NJ, 1989, 451.
3. Cabrera, N., and N.F. Mott. *Rep. Prog. Phys.* 12, 163, 1948–1949.
4. Macdonald, D.D. *Electrochim. Acta* 56, 1761, 2011.
5. Mills, D. J., and S. S. Jamali. *Prog. Org. Coat.* 102, Part A, 8, 2017.
6. J.E.O. Mayne, Pigment electrochemistry. In *Pigment Handbook, Vol. III: Characterisation and Physical Relationships*, ed. T. C. Patton, New York 1973, pp. 457.
7. Mayne, J.E.O., and E.H. Ramshaw. *J. Appl. Chem.* 13, 553, 1969.
8. Troyk, P.R., M.J. Watson, and J.J. Poyezdala. Humidity testing of silicone polymers for corrosion control of implanted medical electronic prostheses. In *Polymeric Materials for Corrosion Control*, ed. R.A. Dickie and F.L. Floyd. Washington, DC: American Chemical Society, 1986, p. 299.

9. European Union. End-of-life vehicles. Directive 2000/53/EC, 2000.
10. del Amo, B., R. Romagnoli, V.F. Vetere, and L.S. Hernández. *Prog. Org. Coat.* 33, 28, 1998.
11. Leidheiser, H. *Prog. Org. Coat.* 7, 79, 1979.
12. Juzeliū nas, E., A. Sudavič ius, K. Jüttner, and W. Fürbeth. *Electrochem. Commun.* 5, 154, 2003.
13. Walter, G.W. *Corros. Sci.* 32, 1041, 1991.
14. Steinsmo, U., and E. Bardal. *Corrosion* 48, 910, 1992.
15. Szillies, S., P. Thissen, D. Tabatabai, F. Feil, W. Fürbeth, N. Fink, and G. Grundmeier. *Appl. Surf. Sci.* 283, 339, 2013.
16. Królikowska, A., *Prog. Org. Coat.* 39, 37, 2000.
17. Koehler, E.L. Corrosion under organic coatings. In *Proceedings of U.R. Evans International Conference on Localized Corrosion*. Houston: NACE International, 1971, p. 117.

3 Generic Types of Anticorrosion Coatings

3.1 COATING COMPOSITION DESIGN

Generally, the formulation of a coating may be said to consist of the binder, pigment, fillers, additives, and carrier (solvent). The binder and the pigment are the most important elements; they may be said to perform the corrosion protection work in the cured paint.

With very few exceptions (e.g., inorganic zinc-rich primers [ZRPs]), binders are organic polymers. A combination of polymers is frequently used, even if the coating belongs to one generic class. An acrylic paint, for example, may purposely use several acrylics derived from different monomers or from similar monomers with varying molecular weights and functional groups of the final polymer. Polymer blends capitalize on each polymer's special characteristics; for example, a methacrylate-based acrylic with its excellent hardness and strength should be blended with a softer polyacrylate to give some flexibility to the cured paint.

Pigments are added for corrosion protection and color. Anticorrosion pigments are chemically active in the cured coating, whereas pigments in barrier coatings must be inert. Fillers must be inert at all times, of course. Coloring pigments of a coating should also stay constant throughout its service life.

Additives may alter certain characteristics of the binder, pigment, or carrier to improve processing and compatibility of the raw materials or application and cure of the coating.

The carrier is the vehicle in the uncured paint that carries the binder, the pigments, and the additives. It exists only in the uncured state. Carriers are liquids in the case of solvent-borne and waterborne coatings, and gases in the case of powder coatings.

3.2 BINDER TYPES

The binder of a cured coating is analogous to the skeleton and skin of the human body. In the manner of a skeleton, the binder provides the physical structure to support and contain the pigments and additives. It binds itself to these components and to the metal surface—hence its name. It also acts somewhat as a skin: the amounts of oxygen, ions, water, and ultraviolet (UV) radiation that can penetrate into the cured coating layer depend to some extent on which polymer is used. The cured coating consists of a very thin polymer-rich or pure-polymer top layer and a heterogeneous mix of pigment particles and binder beneath. The thin topmost layer—sometimes known as the *healed layer* of the coating—covers gaps between pigment particles and cured binder, through which water finds its easiest route to the metal surface. It can also cover pores in the bulk of the coating, blocking this means of water transport. Because this healed surface is very thin, however, its ability to entirely prevent water and oxygen uptake is greatly limited. The ability to absorb, rather than transmit, UV radiation is polymer dependent; acrylics, for example, are for most purposes impervious to UV light, whereas epoxies are extremely sensitive to it.

The binders used in anticorrosion paints are almost exclusively organic polymers. The only commercially significant exceptions are the silicon-based binder in inorganic ZRP siloxanes, and high-temperature silicone coatings. Many of the coating's physical and mechanical properties—including flexibility, hardness, chemical resistances, UV vulnerability, and water and oxygen transport—are determined wholly or in part by the particular polymer or blend of polymers used.

Combinations of monomers and polymers are commonly used, even if the coating belongs to one generic polymer class. Literally hundreds of acrylics are commercially available, all chemically unique; they differ in molecular weights, functional groups, starting monomers, and other characteristics. A paint formulator may purposely blend several acrylics to take advantage of the characteristics of each; thus, a methacrylate-based acrylic with its excellent hardness and strength might be blended with one of the softer polyacrylates to impart flexibility to the cured paint.

Hybrids, or combinations of different polymer families, are also used. Examples of hybrids include acrylic–alkyd hybrid waterborne paints and the epoxy-modified alkyds known as epoxy ester paints.

3.3 EPOXIES

Because of their superior strength, chemical resistance, and adhesion to substrates, epoxies are the most important class of anticorrosive paint. In general, epoxies have the following features:

- Very strong mechanical properties
- Very good adhesion to metal substrates
- Excellent chemical and water resistance
- Better alkali resistance than most other types of polymers
- Susceptibility to UV degradation
- *Can* be sensitive to acids

3.3.1 CHEMISTRY

The term *epoxy* refers to thermosetting polymers produced by reaction of an epoxide group (also known as the glycidyl, epoxy, or oxirane group; see Figure 3.1). The ring structure of the epoxide group provides a site for crosslinking with proton donors, usually amines or polyamides [1].

FIGURE 3.1   Epoxide or oxirane group.

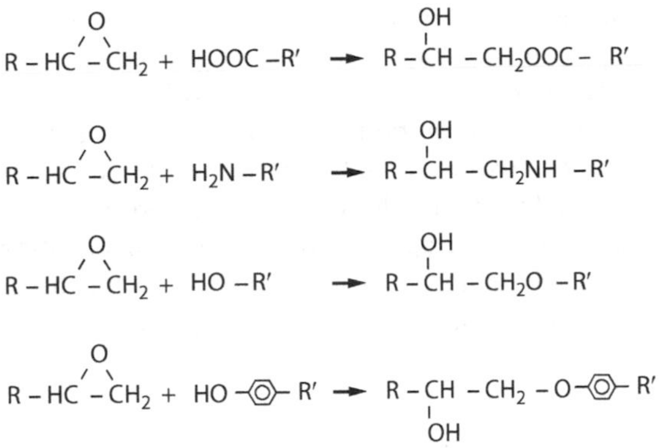

$$R - HC \overset{O}{\underset{\diagdown}{\diagup}} CH_2 + HOOC - R' \longrightarrow R - \overset{OH}{\underset{|}{CH}} - CH_2OOC - R'$$

$$R - HC \overset{O}{\underset{\diagdown}{\diagup}} CH_2 + H_2N - R' \longrightarrow R - \overset{OH}{\underset{|}{CH}} - CH_2NH - R'$$

$$R - HC \overset{O}{\underset{\diagdown}{\diagup}} CH_2 + HO - R' \longrightarrow R - \overset{OH}{\underset{|}{CH}} - CH_2O - R'$$

$$R - HC \overset{O}{\underset{\diagdown}{\diagup}} CH_2 + HO - \bigcirc - R' \longrightarrow R - \underset{\underset{OH}{|}}{CH} - CH_2 - O - \bigcirc - R'$$

FIGURE 3.2 Typical reactions of the epoxide (oxirane) group to form epoxies.

Epoxies have a wide variety of forms, depending on whether the epoxy resin (which contains the epoxide group) reacts with a carboxyl, hydroxyl, phenol, or amine curing agent. Some of the typical reactions and resulting polymers are shown in Figure 3.2. The most commonly used epoxy resins are [2]

- Diglycidyl ethers of bisphenol A (DGEBA or Bis A epoxies)
- Diglycidyl ethers of bisphenol F (DGEBF or Bis F epoxies)—used for low-molecular-weight epoxy coatings
- Epoxy phenol or cresol novolac multifunctional resins

Curing agents include [2]

- Aliphatic polyamines
- Polyamine adducts
- Ketimines
- Polyamides or amidoamines
- Aromatic amines
- Cycloaliphatic amines
- Polyisocyanates

3.3.2 ULTRAVIOLET DEGRADATION

Epoxies are known for their susceptibility to UV degradation. The UV rays of the sun contain enough energy to break certain bonds in the polymeric structure of a cured epoxy binder. As more and more bond breakage occurs in the top surface of the cured binder layer, the polymeric backbone begins to break down. Because the topmost surface or "healed layer" of the cured coating contains only binder, the initial result of the UV degradation is simply loss of gloss. However, as the degradation works downward through the coating layer, binder breakdown begins to free pigment particles. A fine powder consisting of pigment and fragments of binder continually forms on the surface of the coating. The powder is reminiscent of chalk dust—hence the name *chalking* for this breakdown process.

Chalking also occurs to some extent with several other types of polymers. It does not directly affect corrosion protection but is a concern because it eventually results in a thinner coating. Usually, it is considered mainly to be an aesthetic problem as color fades and gloss is lost. The problem is easily overcome with epoxies, however, by covering the epoxy layer with a coating that contains a UV-resistant binder. Polyurethanes are frequently used for this purpose because they are similar in chemical structure to epoxies but are not susceptible to UV breakdown.

3.3.3   VARIETY OF EPOXY PAINTS

The resins used in the epoxy reactions described in Section 3.3.1 are available in a wide range of molecular weights. In general, as molecular weight increases, flexibility, adhesion, substrate wetting, pot life, viscosity, and toughness increase. Increased molecular weight also corresponds to decreased crosslink density, solvent resistance, and chemical resistance [2]. Resins of differing molecular weights are usually blended to provide the balance of properties needed for a particular type of coating.

The number of epoxide reactions possible is practically infinite and has resulted in a huge variety of epoxy polymers. Paint formulators have taken advantage of this variability to provide epoxy paints with a wide range of physical, chemical, and mechanical characteristics. The term *epoxy* encompasses an extremely wide range of coatings, from very low-viscosity epoxy sealers (for penetration of crevices) to exceptionally thick epoxy mastic coatings.

3.3.3.1   Epoxy Mastics

Mastics are high-solid, high-build epoxy coatings designed for situations in which surface preparation is less than ideal. They are sometimes referred to as "surface tolerant" because they provide decent performance on surfaces with less good pretreatment, for example, on grinded or steel brushed surfaces. Due to the low-molecular-weight epoxy in the binder, they have excellent penetrating properties into remaining surface rust and so forth. Mastics can tolerate a lack of surface profile (for anchoring) and a certain amount of contamination that would cause other types of paints to quickly fail.

Formulation is challenging, because the demands placed on this class can be contradictory. Because they are used on smoother and less clean surfaces, mastics must have good wetting characteristics. At the same time, viscosity should be high to prevent sagging of the thick wet film on vertical surfaces.

Epoxy mastics with aluminum flake pigments have very low-moisture permeations and are popular as both spot primers and full coats. Mastics are also used as full-coat primers, usually the variants pigmented with aluminum flake.

Because of their high dry-film thickness, buildup of internal stress in the coating during cure is often an important consideration in using mastic coatings.

3.3.3.2 Solvent-Free Epoxies

Another type of commonly used epoxy paint is the solvent-free, or 100% solid, epoxies. Despite their name, these epoxies are not completely solvent-free. The levels of organic solvents are very low, typically below 5%, which allows very high film builds and greatly reduces concerns about volatile organic compounds (VOCs). An interesting note about these coatings is that many of them generate significant amounts of heat upon mixing. The crosslinking is exothermic, and little solvent is present to take up the heat in vaporization [2]. The wet paint is typically very viscous, and heating may be required during spraying in order to decrease the viscosity. High-pressure spray guns may also be required.

3.3.3.3 Glass Flake Epoxies

Glass flake epoxy coatings are used to protect steel in extremely aggressive environments. When these coatings were first introduced, they were primarily used in offshore applications. In recent years, however, they have been gaining acceptance in mainstream infrastructure as well. Glass flake pigments are wide and thin, and when the paint flows out on the surface during application, they tend to orient themselves parallel to the substrate. This allows them to form layers with a large degree of overlap, creating a highly effective barrier against moisture and chemical penetration because the pathway around and between the glass flakes is extremely long. The glass pigment can also confer increased impact and abrasion resistance and may aid in relieving internal stress in the cured coating. Figure 3.3 shows the cross section of a glass flake epoxy. The horizontal bars of varying thickness are the glass flakes.

3.3.3.4 Epoxy Novolac

Novolacs are phenol-formaldehyde resins. Epoxydized novolacks are produced by reacting the novolac with epichlorhydrin. Epoxy novolac resins contain more than two epoxy groups per molecule and are therefore described as multifunctional epoxy

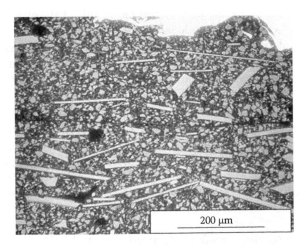

FIGURE 3.3 Cross-sectional picture of a glass flake epoxy.

resins. Due to the high crosslinking density, they have better chemical resistance and temperature resistance than regular epoxies.

3.3.4 HEALTH ISSUES

Components of the uncured epoxy can cause skin allergies, or contact eczema. The reaction is most often seen on hands and underarms, which are the parts of the body that are in contact with the epoxy. The allergy can break out after only short exposure to epoxy and will last for life. The solvent-free and high-solid epoxies contain shorter epoxy polymers in order for them to be liquid. They therefore have a higher vapor pressure, which makes them even more allergenically potent. After developing this allergy, one must stay away from contact with epoxy. Protection is therefore essential when working with epoxy.

3.4 ACRYLICS

Acrylics is a term used to describe a large and varied family of polymers. General characteristics of this group include

- Outstanding UV stability
- Good mechanical properties, particularly toughness [3]

Their exceptional UV resistance makes acrylics particularly suitable for applications in which retention of clarity and color are important.

Acrylic polymers can be used in both waterborne and solvent-borne coating formulations. For anticorrosion paints, the term *acrylic* usually refers to waterborne or latex formulations.

3.4.1 CHEMISTRY

Acrylics are formed by radical polymerization. In this chain of reactions, an initiator—typically a compound with an azo link (—N=N—) or a peroxy link (—0–0—)—breaks down at the central bond, creating two free radicals. These free radicals combine with a monomer, creating a larger free radical molecule, which continues to grow as it combines with monomers, until it either

- Combines with another free radical (effectively canceling each other)
- Reacts with another free radical: briefly meeting, transferring electrons, and splitting unevenly, so that one molecule has an extra hydrogen atom and one is lacking a hydrogen atom (a process known as disproportionation)
- Transfers the free radical to another polymer, a solvent, or a chain transfer agent, such as a low-molecular-weight mercaptan to control molecular weight

This process, excluding transfer, is depicted in Table 3.1 [4].

Some typical initiators used are listed here and shown in Figure 3.4.

TABLE 3.1

Main Reactions Occurring in Free Radical Chain Addition Polymerization

Reaction	Radical Polymerization
Initiator breakdown	$I{:}I \rightarrow I + I$
Initiation and propagation	$I + M_n \rightarrow I(M)_n$
Termination by combination	$I(M)_n + (M)_m I \rightarrow I(M)_{m+n} I$
Termination by disproportionation	$I(M)_n + (M)_m I \rightarrow I(M)_{n-1}(M - H) + I(M)_{m-1}(M + H)$

Source: Data from Bentley, J., Organic film formers, in *Paint and Surface Coatings Theory and Practice*, ed. R. Lambourne, Ellis Horwood Ltd., Chichester, 1987.

Note: I = initiator; M = monomer.

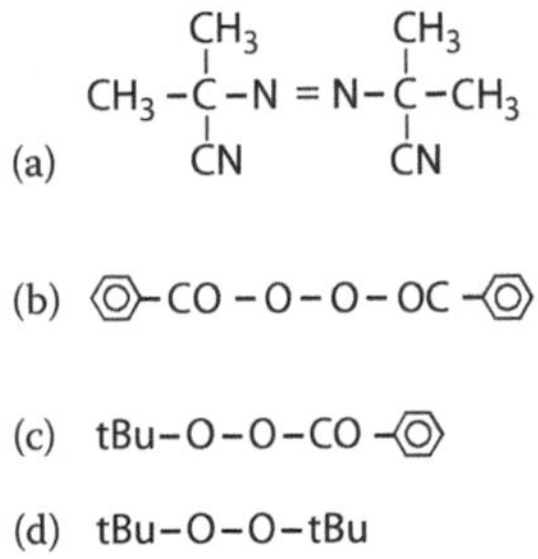

$$\underset{(a)}{\underset{\underset{CN}{|}}{CH_3-\underset{|}{\overset{\overset{CH_3}{|}}{C}}-N=N-\underset{|}{\overset{\overset{CH_3}{|}}{C}}-CH_3}}$$
$$CN \qquad CN$$

(b) ⟨O⟩–CO–O–O–OC–⟨O⟩

(c) tBu–O–O–CO–⟨O⟩

(d) tBu–O–O–tBu

FIGURE 3.4 Typical initiators in radical polymerization: (a) AZDN, (b) dibenzoyl peroxide, (c) *T*-butyl perbenzoate, and (d) di-*t*-butyl peroxide.

- Azo diisobutyronitrile (AZDN)
- Dibenzoyl peroxide
- *T*-butyl perbenzoate
- Di-*t*-butyl peroxide

Typical unsaturated monomers include

- Methacrylic acid
- Methyl methacrylate
- Butyl methacrylate
- Ethyl acrylate
- 2-Ethyl hexyl acrylate
- 2-Hydroxy propyl methacrylate
- Styrene
- Vinyl acetate (Figure 3.5)

3.4.2 SAPONIFICATION

Acrylics can be somewhat sensitive to alkali environments—such as those which can be created by zinc surfaces [5]. This sensitivity is nowhere near as severe as that of alkyds and is easily avoided by proper choice of copolymers.

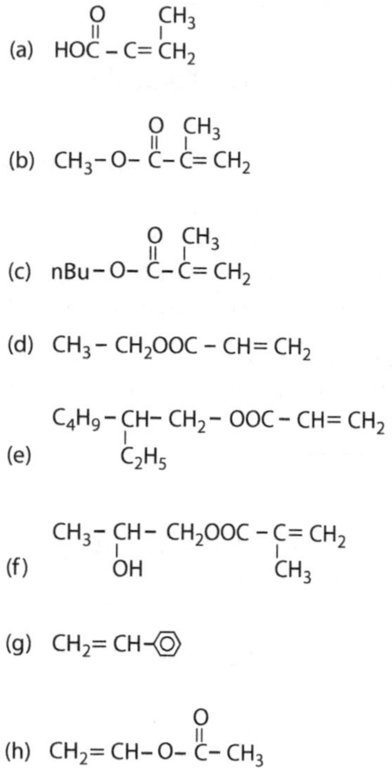

(a) $HO\overset{O}{\overset{\|}{C}} - \overset{CH_3}{\underset{|}{C}} = CH_2$

(b) $CH_3 - O - \overset{O}{\overset{\|}{C}} - \overset{CH_3}{\underset{|}{C}} = CH_2$

(c) $nBu - O - \overset{O}{\overset{\|}{C}} - \overset{CH_3}{\underset{|}{C}} = CH_2$

(d) $CH_3 - CH_2OOC - CH = CH_2$

(e) $C_4H_9 - \underset{\underset{C_2H_5}{|}}{CH} - CH_2 - OOC - CH = CH_2$

(f) $CH_3 - \underset{\underset{OH}{|}}{CH} - CH_2OOC - \underset{\underset{CH_3}{|}}{C} = CH_2$

(g) $CH_2 = CH - \bigcirc$

(h) $CH_2 = CH - O - \overset{O}{\overset{\|}{C}} - CH_3$

FIGURE 3.5 Typical unsaturated monomers: (a) methacrylic acid, (b) methyl methacrylate, (c) butyl methacrylate, (d) ethyl acrylate, (e) 2-ethyl hexyl acrylate, (f) 2-hydroxy propyl methacrylate, (g) styrene, and (h) vinyl acetate.

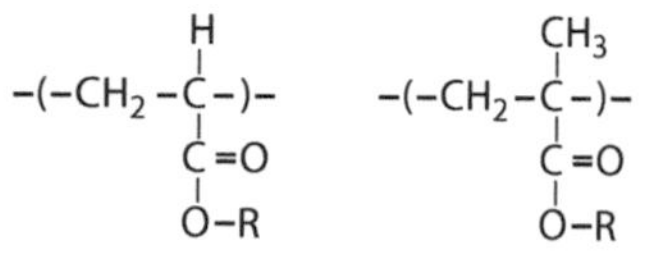

$$-(-CH_2 - \underset{\underset{\underset{\underset{O-R}{|}}{\overset{\|}{C}=O}}{\overset{|}{C}}}{\overset{\overset{H}{|}}{}} -)- \qquad -(-CH_2 - \underset{\underset{\underset{\underset{O-R}{|}}{\overset{\|}{C}=O}}{\overset{|}{C}}}{\overset{\overset{CH_3}{|}}{}} -)-$$

FIGURE 3.6 Depiction of an acrylate (left) and a methacrylate (right) polymer molecule.

Acrylics can be divided into two groups, acrylates and methacrylates, depending on the original monomer from which the polymer was built. As shown in Figure 3.6, the difference lies in a methyl group attached to the backbone of the polymer molecule of a methacrylate in place of the hydrogen atom found in the acrylate.

Poly(methyl methacrylate) is quite resistant to alkaline saponification; the problem lies with the polyacrylates [6]. However, acrylic emulsion polymers cannot be composed solely of methyl methacrylate because the resulting polymer would have a minimum film formation temperature of more than 100°C. Forming a film at room temperature with methyl methacrylate would require unacceptably high amounts of external plasticizers or coalescing solvents. For paint formulations, acrylic emulsion polymers must be copolymerized with acrylate monomers.

TABLE 3.2
Mechanical Properties of Methyl Methacrylate and Polyacrylates

	Methyl Methacrylate	Polyacrylates
Tensile strength (psi)	9000	3–1000
Elongation at break	4%	750%–2000%

Source: Modified from Brendley, W.H., *Paint Varnish Prod.*, 63, 19, 1973.

Acrylics can be successfully formulated for coating zinc or other potentially alkali surfaces, if careful attention is given to the types of monomer used for copolymerization.

3.4.3 COPOLYMERS

Most acrylic coatings are copolymers, in which two or more acrylic polymers are blended to make the binder. This practice combines the advantages of each polymer. Poly(methyl methacrylate), for example, is resistant to saponification, or alkali breakdown. This makes it a highly desirable polymer for coating zinc substrates or any surfaces where alkali conditions may arise. Certain other properties of methyl methacrylate, however, require some modification from a copolymer in order to form a satisfactory paint. For example, the elongation of pure methyl methacrylate is undesirably low for both solvent-borne and waterborne coatings (Table 3.2) [7]. A "softer" acrylate copolymer is therefore used to impart to the binder the necessary ability to flex and bend. Copolymers of acrylates and methacrylates can give the binder the desired balance between hardness and flexibility. Among other properties, acrylates give the coating improved cold crack resistance and adhesion to the substrate, whereas methacrylates contribute toughness and alkali resistance [3,4,6]. In waterborne formulations, methyl methacrylate emulsion polymers alone could not form films at room temperature without high amounts of plasticizers, coalescing solvents, or both.

Copolymerization is also used to improve solvent and water release in the wet stage, and resistance to solvents and water absorption in the cured coating. Styrene is used for hardness and water resistance, and acrylonitrile imparts solvent resistance [3].

3.5 POLYURETHANES

Polyurethanes as a class have the following characteristics:

- Excellent water resistance [1]
- Good resistance to acids and solvents
- Better alkali resistance than most other polymers
- Good abrasion resistance and, in general, good mechanical properties

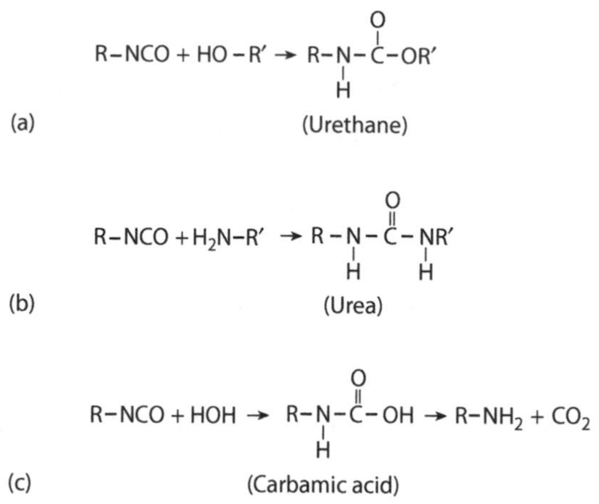

$$R-NCO + HO-R' \rightarrow R-N-\overset{\overset{\displaystyle O}{\|}}{C}-OR'$$
$$\underset{\displaystyle H}{|}$$

(a)   (Urethane)

$$R-NCO + H_2N-R' \rightarrow R-\underset{\underset{\displaystyle H}{|}}{N}-\overset{\overset{\displaystyle O}{\|}}{C}-\underset{\underset{\displaystyle H}{|}}{N}R'$$

(b)   (Urea)

$$R-NCO + HOH \rightarrow R-\underset{\underset{\displaystyle H}{|}}{N}-\overset{\overset{\displaystyle O}{\|}}{C}-OH \rightarrow R-NH_2 + CO_2$$

(c)   (Carbamic acid)

FIGURE 3.7 Some typical isocyanate reactions: (a) hydroxyl reaction, (b) amino reaction, and (c) moisture core reaction.

They are formed by isocyanate (R–N=C=O) reactions, typically with hydroxyl groups, amines, or water. Some typical reactions are shown in Figure 3.7. Polyurethanes are classified into two types, depending on their curing mechanisms: moisture-cure urethanes and chemical-cure urethanes [1]. These are described in more detail in subsequent sections. Both moisture-cure and chemical-cure polyurethanes can be made from either aliphatic or aromatic isocyanates.

Aromatic polyurethanes are made from isocyanates that contain unsaturated carbon rings, for example, toluene diisocyanate (TDI). Aromatic polyurethanes cure faster due to the inherently higher chemical reactivity of the polyisocyanates [8], have more chemical and solvent resistance, and are less expensive than aliphatics but more susceptible to UV radiation [1,9,10]. They are mostly used, therefore, as primers or intermediate coats in conjunction with nonaromatic topcoats that provide UV protection. The UV susceptibility of aromatic polyurethane primers means that the time that elapses between applying coats is very important. The manufacturer's recommendations for maximum recoat time should be carefully followed.

Aliphatic polyurethanes are made from isocyanates that do not contain unsaturated carbon rings. They may have linear or cyclic structures; in cyclic structures, the ring is saturated [11]. The UV resistance of aliphatic polyurethanes is higher than that of aromatic polyurethanes, which results in better weathering characteristics, such as gloss and color retention. For outdoor applications in which good weatherability is necessary, aliphatic topcoats are preferable [1,9]. In aromatic–aliphatic blends, even small amounts of an aromatic component can significantly affect gloss retention [12].

3.5.1 MOISTURE-CURE URETHANES

Moisture-cure urethanes are one-component coatings. The resin has at least two isocyanate groups (–N=C=O) attached to the polymer. These functional groups react with anything containing reactive hydrogen, including water, alcohols, amines, ureas, and other polyurethanes. In moisture-cure urethane coatings, some of the isocyanate reacts with water in the air to form carbamic acid, which is unstable. This acid decomposes to an amine that, in turn, reacts with other isocyanates to form a urea. The urea can continue reacting with any available isocyanates, forming a biuret structure, until all the reactive groups have been consumed [9,11]. Because each molecule contains at least two $-N=C=O$ groups, the result is a crosslinked film.

Because of their curing mechanism, moisture-cure urethanes are tolerant of damp surfaces. Too much moisture on the substrate surface is, of course, detrimental, because isocyanate reacts more easily with water rather than with reactive hydrogen on the substrate surface, leading to adhesion problems. Another factor that limits how much water can be tolerated on the substrate surface is carbon dioxide (CO_2). CO_2 is a product of isocyanate's reaction with water. Too-rapid CO_2 production can lead to bubbling, pinholes, or voids in the coating [9].

Pigmenting moisture-cure polyurethanes is not easy because, like all additives, pigments must be free from moisture [9]. The color range is therefore somewhat limited compared with the color range of other types of coatings.

3.5.2 CHEMICAL-CURE URETHANES

Chemical-cure urethanes are two-component coatings, with a limited pot life after mixing. The reactants in chemical-cure urethanes are

- A material containing an isocyanate group ($-N=C=O$)
- A substance bearing free or latent active hydrogen-containing groups (i.e., hydroxyl or amino groups) [8]

The first reactant acts as the curing agent. Five major monomeric diisocyanates are commercially available [10]:

1. TDI
2. Methylene diphenyl diisocyanate (MDI)
3. Hexamethylene diisocyanate (HDI)
4. Isophorone diisocyanate (IPDI)
5. Hydrogenated MDI ($H_{12}MDI$)

The second reactant is usually a hydroxyl group–containing oligomer from the acrylic, epoxy, polyester, polyether, or vinyl classes. Furthermore, for each of the aforementioned oligomer classes, the type, molecular weight, number of crosslinking sites, and glass transition temperature of the oligomer affect the performance of the coating. This results in a wide range of properties possible in each class of polyurethane coating. The performance ranges of the different types of urethanes overlap, but some broad generalization is possible. Acrylic urethanes, for example, tend to have superior resistance to sunlight, whereas polyester urethanes have better chemical resistance [1,10]. Polyurethane coatings containing polyether polyols generally have better hydrolysis resistance than acrylic- or polyester-based polyurethanes [10].

It should be emphasized that these are very broad generalizations; the performance of any specific coating depends on the particular formulation. It is entirely possible, for example, to formulate polyester polyurethanes that have excellent weathering characteristics.

The stoichiometric balance of the two reactants affects the final coating performance. Too little isocyanate can result in a soft film, with diminished chemical and weathering resistance. A slight excess of isocyanate is not generally a problem, because extra isocyanate can react with the trace amounts of moisture usually present in other components, such as pigments and solvents, or can react over time with ambient moisture. This reaction of excess isocyanate forms additional urea groups, which tend to improve film hardness. Too much excess isocyanate, however, can make the coating harder than desired, with a decrease in impact resistance. Bassner and Hegedus report that isocyanate/polyol ratios (NCO/OH) of 1.05 to 1.2 are commonly used in coating formulations to ensure that all polyol is reacted [11]. Unreacted polyol can plasticize the film, reducing hardness and chemical resistance.

3.5.3 BLOCKED POLYISOCYANATES

An interesting variation of urethane technology is that of the blocked polyisocyanates. These are used when chemical-cure urethane chemistry is desired but, for technical or economical reasons, a two-pack coating is not an option. Heat is needed for deblocking the isocyanate, so these coatings are suitable for use in workshops and plants, rather than in the field.

Creation of the general chemical composition consists of two steps:

1. Heat is used to deblock the isocyanate.
2. The isocyanate crosslinks with the hydrogen-containing coreactant (Figure 3.8).

An example of the application of blocked polyisocyanate technology is polyurethane powder coatings. These coatings typically consist of a solid, blocked isocyanate and a solid polyester resin melt blended with pigments and additives, extruded, and then pulverized. The block polyisocyanate technique can also be used to formulate waterborne polyurethane coatings [8].

Additional information on the chemistry of blocked polyisocyanates is available in reviews by Potter et al. [13] and Wicks [14–16].

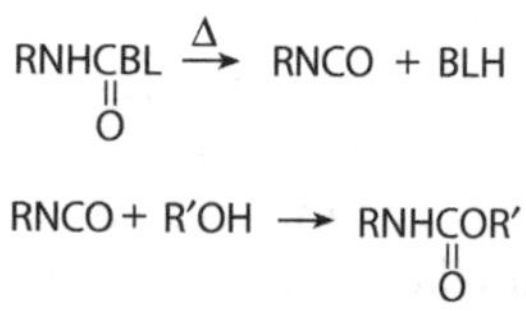

$$RNHCBL \xrightarrow{\Delta} RNCO + BLH$$
$$\underset{\|}{O}$$

$$RNCO + R'OH \longrightarrow RNHCOR'$$
$$\underset{\|}{O}$$

FIGURE 3.8 General reaction for blocked isocyanates.

3.5.4 HEALTH ISSUES

Overexposure to polyisocyanates can irritate the eyes, nose, throat, skin, and lungs. It can cause lung damage and a reduction in lung function. Skin and respiratory sensitization resulting from overexposure can result in asthmatic symptoms that may be permanent. Workers must be properly protected when mixing and applying polyurethanes, as well as when cleaning up after paint application. Inhalation, skin contact, and eye contact must be avoided. The polyurethane coating supplier should be consulted about appropriate personal protective equipment for the formulation. Diisocyanates may be released during thermal degradation of polyurethanes, for example, grinding or welding on polyurethane-coated surfaces. Workers should also protect themselves during such operations.

3.5.5 WATERBORNE POLYURETHANES

For many years, it was thought that urethane technology could not effectively be used for waterborne systems because isocyanates react with water. In the past 20 years, however, waterborne polyurethane technology has evolved tremendously, and in the past few years, two-component waterborne polyurethane systems have achieved some commercial significance.

For information on the chemistry of two-component waterborne polyurethane technology, the reader should see the review of Wicks et al. [16]. A very good review of the effects of two-component waterborne polyurethane formulation on coating properties and application is available from Bassner and Hegedus [11].

3.6 POLYESTERS

Polyester and vinyl ester coatings have been used since the 1960s. Their characteristics include

- Good solvent and chemical resistance, especially acid resistance (polyesters often maintain good chemical resistance at elevated temperatures [17])
- Vulnerability to attack of the ester linkage under strongly alkaline conditions

Because polyesters can be formulated to tolerate very thick film builds, they are popular for lining applications. As thin coatings, they are commonly used for coil-coated products. Many powder coatings are also based on polyesters (see Chapter 6).

3.6.1 CHEMISTRY

Polyester is a very broad term that encompasses both thermoplastic and thermosetting polymers. In paint formulations, only thermosetting polyesters are used. Polyesters used in coatings are formed through

- Condensation of an alcohol and an organic acid, forming an ester. This is the unsaturated polyester prepolymer. It is dissolved in an unsaturated monomer (usually styrene or a similar vinyl-type monomer) to form a resin.

- Crosslinking of the polyester prepolymer using the unsaturated monomer. A peroxide catalyst is added to the resin so that a free radical addition reaction can occur, transforming the liquid resin into a solid film [17].

A wide variety of polyesters are possible, depending on the reactants chosen. The most commonly used organic acids are isophthalic acid, orthophthalic anhydride, terephthalic acid, fumaric acid, and maleic acid. Alcohol reactants used in condensation include bisphenol A, neopentyl glycol, and propylene glycol [17]. The combinations of alcohol and organic acids used determine the mechanical and chemical properties, thermal stability, and other characteristics of polyesters.

3.6.2 SAPONIFICATION

In an alkali environment, the ester links in a polyester can undergo hydrolysis—that is, the bond breaks and reforms into alcohol and acid. This reaction is not favored in acidic or neutral environments but is favored in alkali environments because the alkali forms a salt with the acid component of the ester. These fatty acid salts are called *soaps*, and hence this form of polymer degradation is known as *saponification*.

The extent to which a particular polyester is vulnerable to alkali attack depends on the combination of reactants used to form the polyester prepolymer and the unsaturated monomer with which it is crosslinked. Many polyesters are therefore resistant to alkali as well and can be applied on concrete.

3.6.3 FILLERS

Fillers are very important in polyester coatings because these resins are unusually prone to buildup of internal stresses. The stresses in cured paint films arise for two reasons: shrinkage during cure and a high coefficient of thermal expansion.

During cure, polyester resins typically shrink a relatively high amount, 8–10 vol% [17]. Once the curing film has formed multiple bonds to the substrate, however, shrinkage can freely occur only in the direction perpendicular to the substrate. Shrinkage is hindered in the other two directions (parallel to the surface of the substrate), thus creating internal stress in the curing film.

Stresses also arise in polyesters due to their high coefficients of thermal expansion. Values for polyesters are in the range of 36 to 72 $\times$ 10^{-6} mm/mm/°C, whereas those for steel are typically only 11 $\times$ 10^{-6} mm/mm/°C [17].

Fillers and reinforcements, typically glass flakes, are important for minimizing the stresses and brittleness in the film. For the same reason, proper surface preparation is essential, and grit blasting to Sa2½ and medium to coarse roughness is recommended.

3.7 ALKYDS

In commercial use since 1927 [18], alkyd resins are among the most widely used anticorrosion coatings. They are one-component air-curing paints and, therefore, are fairly easy to use. Alkyds are relatively inexpensive and can be formulated into both solvent-borne and waterborne coatings.

Alkyd paints are not without disadvantages:

- After cure, they continue to react with oxygen in the atmosphere, creating additional crosslinking and then brittleness as the coating ages [18].
- Alkyds cannot tolerate alkali conditions; therefore, they are unsuitable for zinc surfaces or any surfaces where an alkali condition can be expected to occur, such as concrete.
- They are somewhat susceptible to UV radiation, depending on the specific resin composition [18].
- They are not suitable for immersion service because they lose adhesion to the substrate during immersion in water [18].

In addition, it should be noted that alkyd resins generally exhibit poor barrier properties against moisture vapor. Choosing an effective anticorrosion pigment is therefore important for this class of coating [1].

3.7.1 CHEMISTRY

Alkyds are a form of polyester. The main acid ingredient in an alkyd is phthalic acid or its anhydride, and the main alcohol is usually glycerol [18]. Through a condensation reaction, the organic acid and the alcohol form an ester. When the reactants contain multiple alcohol and acid groups, a crosslinked polymer results from the condensation reactions [18].

3.7.2 SAPONIFICATION

In an alkali environment, the ester links in an alkyd break down and reform into alcohol and acid. The known propensity of alkyd coatings to saponify makes them unsuitable for use in alkaline environments or over alkaline surfaces. Concrete, for example, is initially highly alkaline, whereas certain metals, such as zinc, become alkaline over time due to their corrosion products.

This property of alkyds should also be taken into account when choosing pigments for the coating. Alkaline pigments such as red lead or zinc oxide can usefully react with unreacted acid groups in the alkyd, strengthening the film; however, this can also create shelf life problems, if the coating gels before it can be applied.

3.7.3 IMMERSION BEHAVIOR

In making an alkyd resin, an excess of the alcohol reagent is commonly used, for reasons of viscosity control. Because alcohols are water soluble, this excess alcohol means that the coating contains water-soluble material and therefore tends to absorb water and swell [18]. Therefore, alkyd coatings tend to lose chemical adhesion to the substrates when immersed in water. This process is usually reversible. As Byrnes describes it, "They behave as if they were attached to the substrate by water-soluble glue" [18]. Alkyd coatings are therefore not suitable for immersion service.

3.7.4 BRITTLENESS

Alkyds cure through a reaction of the unsaturated fatty acid component with oxygen in the atmosphere. Once the coating has dried, the reaction does not stop but continues to crosslink. Eventually, this leads to undesirable brittleness as the coating ages, leaving the coating more vulnerable to, for example, freeze–thaw stresses.

3.7.5 DARKNESS DEGRADATION

Byrnes notes an interesting phenomenon in some alkyds: if left in the dark for a long time, they become soft and sticky. This reaction is most commonly seen in alkyds with high linseed oil content [18]. The reason why light is necessary for maintaining the cured film is not clear.

3.8 POLYSILOXANES

The first polysiloxane-based coatings were introduced on the market in the 1990s, and are now recognized for their combination of protective and decorative properties. The siloxane backbone of the polymer gives a very UV-resistant coating with excellent gloss retention properties. The properties offered by polysiloxane coatings are [19]

- Excellent UV resistance and gloss retention
- Can be formulated to also be a barrier coating
- Good adhesion and cohesion properties
- Good chemical resistance
- High-solid, low-VOC content
- Limited surface tolerance, usually applied with a primer

3.8.1 CHEMISTRY

Polysiloxane is an organic–inorganic hybrid polymer, where the inorganic siloxane resin is reacted with acrylic or epoxy resins. Figure 3.9 shows schematically the structure of an epoxy-modified polysiloxane. Polysilosanes can be formulated as both single component and two component. Curing of single-component polysiloxanes requires the presence of humidity in the atmosphere, because the curing reaction involves hydrolysis of the ethoxy group in the silane, releasing ethanol and producing the Si-OH bond that polymerize and form the silicon polymer. In the two-component polysiloxanes, the silicon chain is already formed in the resin (component A), and the curing is achieved by a silane with the desired functional group in component B. For example, the epoxy silane resin in Figure 3.9 would be cured with an aminosilane (a silane with an amine functional group). The epoxy silane and the aminosilane react in the usual manner, just like in traditional epoxies.

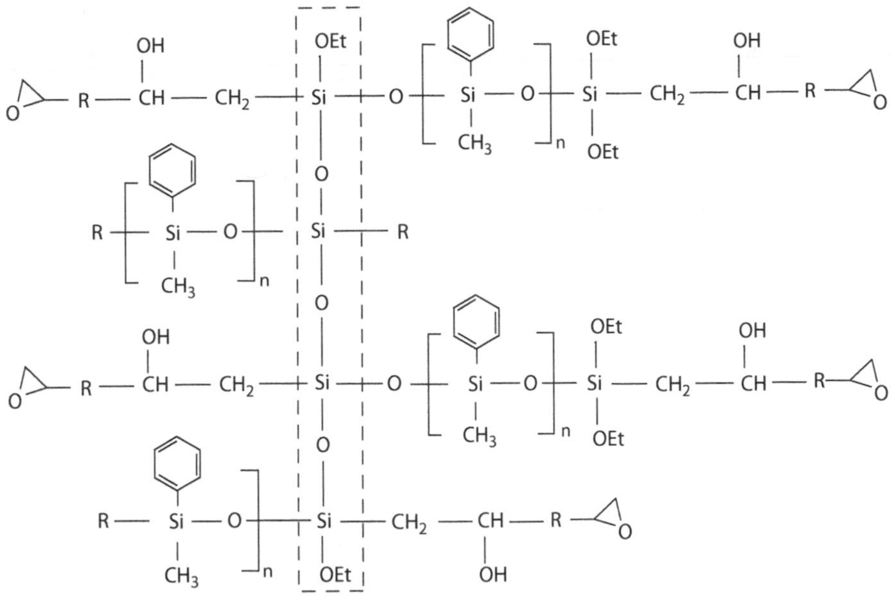

FIGURE 3.9 Structure of epoxy polysiloxane.

The inorganic ratio of the binder varies from 37% to 77% (weight by weight) in commercial products, but is typically around 60% [20]. Formulation with a high crosslink density will result in a strong, hard, protective, and chemically resistant coating, but will also be quite brittle and have low surface tolerance. A higher organic fraction will result in lower crosslink density, which will make the coating more flexible, but not as good a barrier. The latter will be more suitable as a topcoat to be applied, for example, on a protective epoxy, while the former will be UV-resistant barrier coating. Epoxies are known for their poor UV resistance, which indicates that epoxy-modified polysiloxanes will be more susceptible to UV degradation than acrylic polysiloxanes. However, this is not necessarily the case. The amount of organic modification seems to be more important, but even the polysiloxanes with the highest level of organic modification show better gloss retention than polyurethanes [19].

3.8.2 PERFORMANCE OF POLYSILOXANE COATING SYSTEMS

As stated above, polysiloxanes cannot be applied directly on steel, and a primer is required. When formulated as a protective barrier coating, and given the very good UV resistance, they can be used in two-coat systems, for example, applied on top of a zinc epoxy. Two-coat systems have low application costs, and were therefore selected as the main protection system on several offshore platforms. However, this resulted in severe degradation [21]. A thorough investigation of coating properties did not reveal specific weaknesses of the polysiloxane [22–24]. Instead, it was assumed that the failure was due to low film thickness. When painting a large construction like an offshore platform, there will inevitably be variations in film thickness in every

coat, and in some areas the film will be too thin. In a three-coat system, too-low film
thickness in one coat may be compensated by the next coat. In a two-coat system,
this redundancy will be absent, making the system more vulnerable. The NORSOK
M-501 coating specification was later changed due to these experiences, and three-
coat systems are now required [25].

Polysiloxanes are now mainly used as a high-gloss alternative to polyurethanes,
and where there are restrictions to the use of polyurethanes.

3.9 OTHER BINDERS

Other types of binders include epoxy esters and silicon-based inorganic zinc-rich
coatings.

3.9.1 Epoxy Esters

Despite their name, epoxy esters are not really epoxies. Appleman, in fact, writes that
epoxy esters "are best described as an epoxy-modified alkyd" [26]. They are made
by mixing an epoxy resin with either an oil (drying or vegetable) or a drying oil acid.
The epoxy resin does not crosslink in the manner of conventional epoxies. Instead,
the resin and oil or drying oil acid are subjected to high temperature, 240°C–260°C,
and an inert atmosphere to induce an esterification reaction. The result is a binder
that cures by oxidation and can therefore be formulated into one-component paints.

Epoxy esters generally possess adhesion, chemical and UV resistance, and corrosion protection properties that are somewhere between those of alkyds and epoxies
[27]. They also exhibit resistance to splashing of gasoline and other petroleum fuels
and are therefore commonly used to paint machinery [28].

3.9.2 Silicate-Based Inorganic Zinc-Rich Coatings

Silicon-based inorganic zinc-rich coatings are almost entirely zinc pigment; zinc
levels of 90% or higher are common. They contain only enough binder to keep the
zinc particles in electrical contact with the substrate and each other. The binder in
inorganic ZRPs is an inorganic silicate, which may be either a solvent-based, partly
hydrolyzed alkyl silicate (typically ethyl silicate) or a water-based, highly alkali
silicate.

General characteristics of these coatings are

* Ability to tolerate higher temperatures than organic coatings (inorganic
  ZRPs typically tolerate 700°F–750°F)
* Excellent corrosion protection
* Require topcoatings in high-pH or low-pH conditions
* Require a very thorough abrasive cleaning of the steel substrate, typically
  near-white metal (Sa 2 ½)

For a more-detailed discussion of inorganic ZRPs, see Section 4.1.

REFERENCES

1. Smith, L.M. *J. Prot. Coat. Linings* 13, 73, 1995.
2. Salem, L.S. *J. Prot. Coat. Linings* 13, 77, 1996.
3. Flynn, R., and D. Watson. *J. Prot. Coat. Linings* 12, 81, 1995.
4. Bentley, J. Organic film formers. In *Paint and Surface Coatings Theory and Practice*, ed. R. Lambourne. Chichester: Ellis Horwood Limited, 1987.
5. Forsgren, A., M. Linder, and N. Steihed. Substrate-polymer compatibility for various waterborne paint resins. Report 1999:1E. Stockholm: Swedish Corrosion Institute, 1999.
6. Billmeyer, F.W. *Textbook of Polymer Science*. 3rd ed. New York: John Wiley & Sons, 1984, p. 388.
7. Brendley, W.H. *Paint Varnish Prod.* 63, 19, 1973.
8. Potter, T.A., and J.L. Williams. *J. Coat. Technol.* 59, 63, 1987.
9. Gardner, G. *J. Prot. Coat. Linings* 13, 81, 1996.
10. Roesler, R.R., and P.R. Hergenrother. *J. Prot. Coat. Linings* 13, 83, 1996.
11. Bassner, S.L., and C.R. Hegedus. *J. Prot. Coat. Linings* 13, 52, 1996.
12. Luthra, S., and R. Hergenrother. *J. Prot. Coat. Linings* 10, 31, 1993.
13. Potter, T.A., J.W. Rosthauser, and H.G. Schmelzer. In *Proceedings of the 11th International Conference on Organic Coatings Science and Technology*, Athens, 1985, paper 331.
14. Wicks, Z.W., Jr. *Prog. Org. Coat.* 9, 3, 1981.
15. Wicks, Z.W., Jr. *Prog. Org. Coat.* 3, 73, 1975.
16. Wicks, Z.W., Jr., D.A. Wicks, and J.W. Rosthauser. *Prog. Org. Coat.* 44, 161, 2002.
17. Slama, W.R. *J. Prot. Coat. Linings* 13, 88, 1996.
18. Byrnes, G. *J. Prot. Coat. Linings* 13, 73, 1996.
19. Graversen, E. Comparison between epoxy polysiloxane and acrylic polysiloxane finishes. Presented at CORROSION/2007. Houston: NACE International, 2007, paper 7008.
20. Andrews, A.F. Polysiloxane topcoats—A step too far? Presented at CORROSION/2005. Houston: NACE International, 2005, paper 5007.
21. Knudsen, O.Ø. Review of coating failure incidents on the Norwegian continental shelf since the introduction of NORSOK M-501. Presented at CORROSION 2013. Houston: NACE International, 2013, paper 2500.
22. Axelsen, S.B., R. Johnsen, and O. Knudsen. *CORROSION* 66, 125003, 2010.
23. Axelsen, S.B., R. Johnsen, and O. Knudsen. *CORROSION* 66, 065005, 2010.
24. Axelsen, S.B., R. Johnsen, and O.Ø. Knudsen. *CORROSION* 66, 065006, 2010.
25. NORSOK M-501. Surface preparation and protective coatings. Rev. 6. Oslo: Norwegian Technology Standards Institution, 2012.
26. Appleman, B.R. *Corrosioneering* 1, 4, 2001.
27. Kaminski, W. *J. Prot. Coat. Linings* 13, 57, 1996.
28. Hare, C.H. *J. Prot. Coat. Linings* 12, 41, 1995.

4 Corrosion-Protective Pigments

Protective pigments come in three major types: sacrificial, inhibitive, and barrier. Sacrificial pigments are added in large quantities to allow the flow of electric current. When in electrical contact with the steel surface, the sacrificial pigments act as the anode of a large corrosion cell and protect the steel cathode. Coatings utilizing inhibitive pigments release a soluble species, such as molybdates or phosphates, from the pigment into any water that penetrates the coating. These species are carried to the metal surface, where they inhibit corrosion by encouraging the growth of protective surface layers [1] or modify pH in a beneficial manner. Solubility and reactivity are critical parameters for inhibitive pigments; a great deal of research is occupied with controlling the former and decreasing the latter. Both inhibitive and sacrificial pigments are effective only in the layer immediately adjacent to the steel (i.e., the primer). Barrier coatings are probably the oldest type of coating [1], and the requirements of their pigments are completely different. Specifically, chemical inertness and a flake- or plate-like shape are the requirements of barrier pigments. Unlike inhibitive or sacrificial coatings, barrier coatings can be used as primer, intermediate coat, or topcoat because their pigments do not react with metal.

4.1 ZINC DUST

Zinc-rich paints, or zinc-rich primers, (ZRPs) have been used to protect steel construction for many decades [2], and zinc dust is now the most widely used protective pigment in paint. In order for a paint to be called zinc rich, the dry film must contain more than 80% by weight of zinc dust. Some products even contain more than 90% zinc dust. ZRPs come with two groups of binders: inorganic silicate binders and organic binders, typically epoxies. They have different properties and are used in different ways. An important function of both is the cathodic protection effect. The zinc particles corrode and polarize the steel cathodically. In order for this effect to occur, the zinc must be in electrical contact with the steel substrate; that is, ZRP is always applied directly on the steel. They are therefore usually referred to as zinc-rich primers.

An important term when discussing ZRPs is pigment volume concentration (PVC), that is, the volume of the dry film that is occupied by the pigment. At a certain PVC, the binder is no longer able to wet all the pigments and the film becomes porous. This is called the critical pigment volume concentration (CPVC). Many zinc-rich primers have a zinc loading above the CPVC.

4.1.1 Types of Zinc-Rich Paint

There are two classes of ZRPs, which differ depending on the binder used: organic and inorganic [3]. A family tree of the various types of ZRPs is shown in Figure 4.1. Two-component epoxy amines or amides, epoxy esters, and moisture-cure urethanes are examples of organic binders. Organic binders have a dense character and are protective and electrically insulating; for that reason, they may also protect the zinc particles, preventing them from acting as sacrificial anodes. Most organic ZRPs are formulated with a PVC close to the CPVC in order to have sufficient cohesion in the film. This is a compromise on the expense of the sacrificial properties of the coating. Still, most organic ZRPs are able to provide cathodic protection of the steel substrate.

Inorganic binders are silica based. They can be further divided into two groups: solvent-based partly hydrolyzed alkyl silicate (mostly ethyl silicate) and water-based highly alkaline silicates. Ethyl silicates and alkali silicates will form the same film, but the wet paints are quite different. The alkali silicates are waterborne, while the ethyl silicates are solvent-borne. The curing reactions are also different. Ethyl silicates require a certain air humidity in order to cure in a reaction releasing ethanol (Figure 4.2a and b). The ethyl silicate must be somewhat prehydrolyzed to give proper curing. Ethyl silicate therefore sometimes has a storage stability problem. The released ethanol contributes to the volatile organic compound (VOC) emission. The contribution of water in the curing reaction means that a certain air humidity is required in order for the reaction to occur. The alkali silicates polymerize when the water evaporates from the film in a complex reaction, producing the silicate binder and hydroxide, resulting in a high pH. Addition of various substances can reduce the pH in the film. During exposure, the hydroxide will gradually be washed out. The high pH in the alkali silicates makes them less suitable for topcoating. Unsuccessful curing is the most frequent cause of failure for zinc silicate coatings.

Inorganic ZRPs contain very little organic substances and are therefore used as weldable or shop primers. A shop primer is a thin coating that is applied to protect

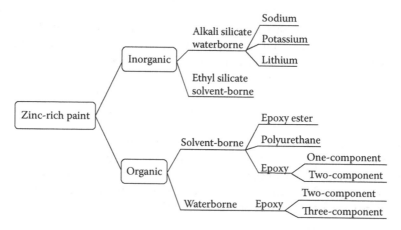

FIGURE 4.1 Family tree of ZRPs. (Data from Undrum, H., *J. Prot. Coat. Linings*, 23, 52, 2006.)

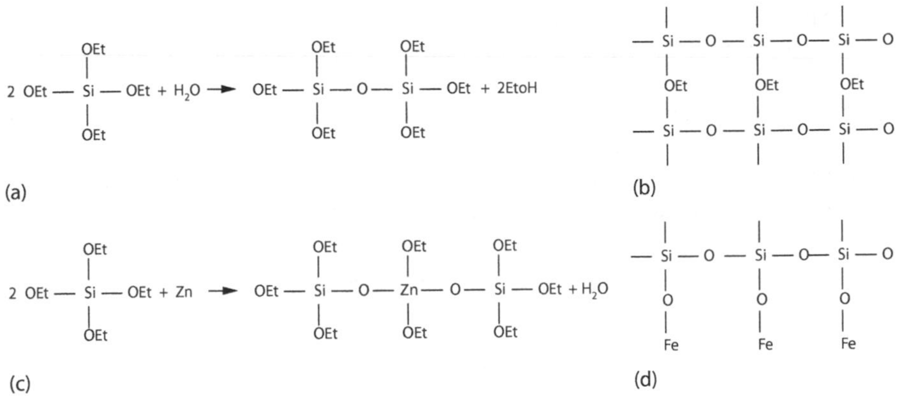

(a)

(b)

(c)

(d)

FIGURE 4.2   Curing reaction of ethyl silicate (a) and resulting silicate binder structure (b). The silicate can react with zinc to form a zinc silicate polymer (c). The silicate can also bond to the oxide covering the steel substrate and the zinc particles (d).

the steel temporarily in the yard during construction [4]. Welding steel with an organic shop primer will produce gases due to oxidation of the binder that have a tendency to enter the weld and create pores, weakening the weld mechanically. By using silicate shop primers, this problem is avoided. The shop primers typically have a rather low zinc loading, typically in the 28%–48% range. Sometimes, the shop primer is kept as a first coat in the final coating system in order to save costs [4]. Permanent zinc silicates are always pigmented above the CPVC and therefore have high porosity. The silicate binder is mechanically stronger than the organic binders are, and can form covalent bonds to both the zinc particles and the steel substrate (Figure 4.2c and d), and sufficient adhesion and cohesion can be obtained with less binder. For this reason, the zinc silicates have a higher capacity for cathodic protection. With time (and corrosion of the zinc), the matrix fills with zinc salts, giving a very dense barrier coat.

Two-component zinc epoxies cure like regular epoxies, which makes the curing much easier to control.

4.1.2   PROTECTION MECHANISMS

Zinc dust is generally recognized to offer corrosion protection to steel via three mechanisms:

1. Cathodic protection to the steel substrate (the zinc acts as a sacrificial anode). This takes place at the beginning of the coating's lifetime and naturally disappears with time [5]. Only metallic zinc will of course have this effect.
2. Barrier action. As a result of the zinc sacrificially corroding, zinc ions are released into the coating. These ions can react with other species in the coating to form insoluble zinc salts. As they precipitate, these salts fill in the pores in the coating, reducing permeability of the film [2].

3. Slightly alkaline conditions are formed as the zinc corrodes, passivating the steel substrate or neutralizing other processes that cause acidification [5].

Of these three mechanisms, the first two depend on a high zinc content to work properly; the last one is independent of zinc content. The cathodic protection mechanism is probably the most important one. In order for the zinc-rich primer to provide cathodic protection, the following premises must be fulfilled:

- As stated above, the zinc must be in electric contact with the substrate.
- The zinc must also be in electrolytic contact with the steel, that is, a water film must be present on the surface.
- The zinc must be active, that is, corroding.

Zinc dust comes in two forms: flake zinc dust and granular grade. Commercial zinc-rich primers are mainly formulated with the granular grade, but flake dust has been shown to improve performance in some investigations [6]. Figure 4.3 shows a cross section of a zinc-rich primer. The particles are in the order of 1–10 μm in diameter. In order for each individual zinc particle to provide cathodic protection of the steel substrate, it must be in electric contact with the steel. The particles close to the substrate may be in direct contact, but for the particles further into the coating, contact can only be achieved through physical contact with particles below, in a chain of particle–particle junctions. If contact in one junction is lost, the particles beyond this junction will be disconnected and unable to contribute to the cathodic protection. Particle size will therefore affect the ability of the coating to provide protection. The smaller particles we have, the more particle–particle junctions there will be for a given film thickness. Hence, large particles then should be beneficial. In fact, a distribution of particle sizes and shapes seems to give the best result, probably due to a combination of the high chance for particle–particle contact and less junctions throughout the film [6]. The cross section in Figure 4.3 indicates that only a few particles are in contact with each other. This does not mean that contact is rare.

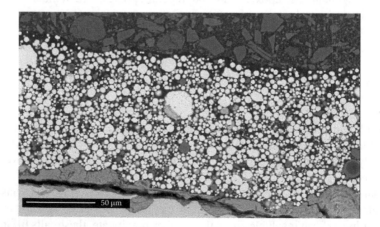

FIGURE 4.3   Cross section of a zinc-rich primer.

The image is a two-dimensional cross section, and there will be more contact points outside this cross section.

Higher zinc loading will also increase the chance for contact throughout the film. However, there is a limit to how much zinc dust can be added to the paint. The formulation is optimized with respect to protective properties and cohesive strength, as discussed above. According to ISO 12944-5, a zinc-rich primer shall contain more than 80% zinc by weight [7]. Zinc epoxies typically contain 80%–85% zinc, while zinc silicates typically contain 85%–90%. Since the zinc has a much higher density than the binder, the volume percentage of zinc is smaller. In a zinc-rich primer with 85% zinc by weight, a zinc epoxy will have about 55% zinc by volume, and a zinc silicate will have about 70% zinc by volume. The latter will normally also be rather porous; that is, not all particles are fully wetted by the silicate. This takes us to the last of the three premises above—the zinc must be corroding actively. The binder may protect the zinc from corrosion, and zinc epoxies in particular may lose much of their capacity for cathodic protection this way. They contain more binder than the zinc silicates, and epoxy is a more protective binder than silicate.

As the zinc particles are sacrificed and corrode, they will be converted to zinc oxide. We may then fear that contact is rapidly lost. Fortunately, zinc oxide is a semiconductor, so even though we may lose the metallic contact between the particles, electrons may still flow through the oxide and protection continues.

4.1.3 Topcoating Zinc-Rich Paint or Not

Some very effective zinc-rich primers allow the conditions necessary for corrosion to occur—water, oxygen, and ions are allowed to penetrate the coating. These coatings do not protect by suppressing the corrosion process, but rely only on the cathodic protection provided by the zinc. This mechanism requires that the zinc-rich primer is not topcoated. This way, small damages in the primer can be protected by the zinc that is available in an area around the damage. The protection potential for steel is –0.80 V (Ag/AgCl), while the potential of the zinc is typically 1.00 V (Ag/AgCl). The potential drop between the zinc and the steel must therefore not be more than 0.20 V. Atmospherically exposed surfaces are only covered by a thin film of electrolyte, which must provide the electrolytic contact between the zinc and the steel. The potential drop in this film will depend on the protection current, its conductivity, and its dimensions, that is, the thickness of the film and the distance between the protecting zinc and the exposed steel. Experiments have shown that full protection is only found a few millimeters from the edge of the zinc coating, but partial protection is found several millimeters away. The film will have a certain capacity for cathodic protection, depending on the film thickness. When all the zinc has corroded, the cathodic protection mechanism is lost. The coating may still protect by displacing water, though. The zinc oxide will fill the pores in the film, giving it some protective abilities still. Zinc epoxies are not suitable for exposure without a topcoat, since ultraviolet (UV) exposure will degrade the epoxy binder [8].

When the zinc-rich primer is coated with one or more protective films, it will be unable to protect even small damages in the coating, since only the very limited amount of zinc that is exposed at the coating damage edge will provide active protection. The zinc that is covered by the protective film is now protected and not in electrolytic contact with the exposed steel. In spite of this, zinc-rich primers are in most cases topcoated. There are mainly two reasons for this. The first is visual appearance. Zinc-rich primers are available in a limited color range, and when they react, they will be covered by white zinc oxides. Hence, the surface will usually have a gray and dull appearance. The second is that long coating lifetimes are also achieved when the primer is overcoated. The coating system will now primarily protect by the water displacement mechanism, but the zinc-rich primer will slow down corrosion creep, that is, the corrosion that creeps under the coating from damages in the film. Corrosion creep is discussed in Chapter 12. The question then is when to topcoat the zinc-rich primer and when not to. In many cases, the color requirements will give the answer. Many constructions have requirements to their visual appearance or signal colors. This cannot be achieved with a zinc-rich primer alone.

4.1.4 Choosing a Zinc-Rich Paint

Zinc epoxies are normally preferred in multilayer coating systems, because they are easier to topcoat. The porous nature of the zinc silicates makes them difficult to paint. The air in the porous zinc silicate cause gas bubbles to form in the subsequent coat, so-called popping. The bubbles often collapse at a stage when the paint is too viscous to float together and close the hole. To avoid this, a tie coat is usually applied on the cured zinc silicate film, typically a diluted epoxy coat applied in a thin film that fills the pores and allows the air to escape without formation of bubbles. This usually solves the problem, but adds one more coat to the system and thereby increases the costs. In addition, the curing of silicates has a reputation for being unreliable [8]. If air humidity is too low, the reaction is slow and curing may take a long time, delaying production in the yard. Diffusion of moisture through the topcoat may be sufficient to cure the silicate in a few weeks, though [8].

When zinc silicates are applied too thick, they have a tendency to crack, so-called mud cracking. When the film cures, tension builds up inside the film and it cracks. The tension increases with film thickness, making zinc silicates very sensitive to high film thickness. This problem has earned them a reputation of being a bit difficult to apply, which has become another argument for choosing zinc epoxies.

The inorganic zinc silicates have a much higher temperature tolerance than the organic zinc epoxies, and can be used at temperatures up to 400°C, that is, approaching the melting point of zinc. Zinc epoxies are limited by the temperature range of the epoxy. The range from 60°C to 100°C should, however, receive some extra consideration before a cathodically protecting coating is applied, due to a chance for potential inversion. The zinc may passify and become a cathode in this temperature range, causing corrosion on the steel [9]. This was reported for hot-dip galvanized steel inside hot water tanks containing hard water. However, the problem has never been reported for atmospherically exposed zinc. In addition, the large surface area of zinc in zinc-rich primers makes this seem unlikely. In fact, the problem is rather the opposite, that is, rapid corrosion of the zinc at elevated temperature. Zinc silicates have been used for corrosion protection of thermally insulated steel, for example, pipes and tanks in process plants. If water enters the insulation, rapid loss of the zinc silicate has been reported [10].

Table 4.1 compares various properties of zinc silicates and zinc epoxy.

TABLE 4.1
Zinc Silicate vs. Zinc Epoxy: Advantages and Disadvantages

	Zinc Ethyl Silicates	Zinc Alkali Silicates	Zinc Epoxy
Topcoating	Good	Not recommended	Excellent
Surface preparation requirements	Sa2½	Sa2½	Sa2½
Temperature range	400°C	400°C	120°C
Cathodic protection	Excellent	Excellent	Good
Ease of application	Fair	Good	Very good
Mud cracking	Fair	Problem at high DFT	Very good
Flexibility	Limited	Limited	Very good
Cohesion	Fair	Fair	Good

Note: DFT = dry-film thickness.

4.2 PHOSPHATES

Phosphates is a term that is used to refer to a large group of pigments that contain a phosphorus and an oxygen functional group. Its meaning is vast: the term *zinc phosphates* alone includes, but is not limited to,

- Zinc phosphate, first-generation $Zn_3(PO_4)_2·4H_2O$
- Aluminum zinc phosphate [11] or zinc aluminum phosphate [12]
- Zinc molybdenum phosphate
- Aluminum zinc hydroxyphosphate [12]
- Zinc hydroxymolybdate phosphate or basic zinc molybdate phosphate [12,13]
- Basic zinc phosphate $Zn_2(OH)PO_4·2H_2O$ [12,13]
- Zinc silicophosphate [14]
- Zinc aluminum polyphosphate [12]

Zinc-free phosphates include

- Aluminum phosphate
- Dihydrogen tripolyphosphates [13]
- Dihydrogen aluminum triphosphate [11,13–15]
- Strontium aluminum polyphosphate [12]
- Calcium aluminum polyphosphate silicate [12]
- Zinc calcium strontium polyphosphate silicate [12]
- Laurylammonium phosphate [16]
- Hydroxyphosphates of iron, barium, chromium, cadmium, and magnesium, for example, $FePO_4·2H_2O$, $Ca_3(PO_4)_2$-1/2H_2O, $Ba_3(PO_4)_2$, $BaHPO_4$, and $FeNH_4PO_4·2H_2O$ [11]

In this section, the pigments discussed in more detail include the zinc phosphates and one type of nonzinc phosphate, aluminum triphosphates.

4.2.1 ZINC PHOSPHATES

Zinc phosphates are widely used in many binders, including oil-based binders, alkyds, and epoxies [17–25]. Their low solubility and activity make them extremely versatile; they can be used in resins, such as alkyds, where more alkali pigments pose stability problems. Typical loading levels are 10%–30% in protective coatings.

The popularity of the zinc phosphate family of pigments is easily understood when the toxicological data are examined. Lead, chromium, barium, and strontium are all labeled toxic in one form or another. Zinc phosphates, however, pose no known chronic toxicity [12].

Zinc phosphates can provide corrosion protection to steel through multiple mechanisms:

- *Phosphate ion donation*: Phosphate ion donation can be used for ferrous metals only [11,13,14,21,26]. As water penetrates through the coating, slight hydrolysis of zinc phosphate occurs, resulting in secondary phosphate ions. These phosphate ions in turn form a protective passive layer [27,28] that, when sufficiently thick, prevents anodic corrosion [29]. Porosity of the phosphate coatings is closely related to the coating protective performance [11]. The approximate formula for the phosphatized metallic compound is $Zn_5Fe(PO_4)_2 \cdot 4H_2O$ [30].
- *Creation of protective films on the anode*: In this model, suggested by Pryor and others [31,32], oxygen dissolved in the film is adsorbed onto the metal. There it undergoes a heterogeneous reaction to form a protective film of γ-Fe_2O_3; this film thickens until it reaches an equilibrium value of 20 nm. The film prevents the outward diffusion of iron. Phosphate ions do not appear to directly contribute to the oxide film formation, but rather act to complete or maintain it by plugging discontinuities with anion precipitates of Fe(III) ions. Romagnoli and Vetere have noted that Pryor used soluble phosphates rather than the generally insoluble phosphates used in coatings, so care should be taken in extrapolating these results [11]. Other studies have found both oxyhydroxides and iron phosphates incorporated in the protective film [33].
- *Inhibitive aqueous extracts formed with certain oleoresinous binders*: Components of the binder, such as carboxylic and hydroxyl groups, form complexes with either the zinc phosphate or the intermediate compounds formed when the zinc phosphate becomes hydrated and dissociates. These complexes can then react with corrosion products to form a tightly adhering, inhibitive layer on the substrate [13,18–21,26,34].
- *Polarization of the substrate*: Clay and Cox [35] have suggested that nearly insoluble basic salts are formed and adhere well to the metal surface. These salts limit the access of dissolved oxygen to the metal surface and polarize

the cathodic areas. This theory was supported by the work of Szklarska-Smialowska and Mankowsky [36].

4.2.2 TYPES OF ZINC PHOSPHATES

Because so many variations of zinc phosphates are available, it is convenient to divide them into groups for discussion. Although no formal classes of zinc phosphates exist, they have been divided here into groups or generations, more or less by chronological development.

The simplest, or first-generation, zinc phosphate is made by either mixing disodium phosphate and zinc sulfate solutions at boiling temperature or saturating a 68% phosphoric acid solution with zinc oxide, also at boiling temperature. Both methods give a precipitate with an extremely coarse crystalline structure. Further treatment yields $Zn_3(PO_4)_2 \cdot 4H_2O$, first-generation zinc phosphate [11]. The usefulness of first-generation zinc phosphate is limited by its low solubility [37]; only a small concentration of phosphate ions is available to protect the metal. This is a problem because corrosion inhibition by phosphates takes place only when the anion concentration is higher than 0.001 M in a salt solution at pH 5.5–7.0 [32].

Zinc phosphates can be modified to increase their solubility in water or to add other functional groups that can also act as inhibitors. This is usually achieved by adding an organic surface treatment to the pigment or blending other inorganic inhibitors with the zinc phosphate [14]. Table 4.2 shows the amount of phosphate ions in milligrams per liter of water obtained from various first-generation and subsequent-generation zinc phosphates [38]. It can be clearly seen why modifying phosphate pigments is an area of great interest: aluminum zinc phosphate provides 250 times the amount of dissolved phosphate as first-generation zinc phosphate.

Second-generation zinc phosphates can be divided into three groups: basic zinc phosphates, salts of phosphoric acid and metallic cations, and orthophosphates.

TABLE 4.2
Relative Solubilities in Water of Zinc Phosphate and Modified Zinc Phosphate Pigments

	Water-Soluble Matter (mg/L) (ASTM D2448-73, 10 g Pigment in 90 ml Water)			
Pigment	Total	Zn^{2+}	PO_4^{3-}	MoO_4^{2-}
Zinc phosphate	40	5	1	—
Organic modified zinc phosphate	300	80	1	—
Aluminum zinc phosphate	400	80	250	—
Zinc molybdenum phosphate	200	40	0.3	17

Source: Bittner, A., *J. Coat. Technol.*, 61, 111, 1989, Table 2. With permission.

First-generation zinc phosphate, $Zn_3(PO_4)_2 \cdot 4H_2O$, is a neutral salt. Basic zinc phosphate, $Zn_2(OH)PO_4 \cdot 2H_2O$, yields a different ratio of Zn^{2+} and PO_4^{3-} ions in solution and has a higher activity than the neutral salt [13]. It has been reported that basic zinc phosphate is as effective a corrosion inhibitor as zinc phosphate plus a mixture of pigments containing water-soluble chromates [39–41].

Another group of second-generation phosphate pigments includes salts formed between phosphoric acid and different metallic cations, for example, hydrated modified aluminum zinc hydroxyphosphate and hydrated zinc hydroxymolybdate phosphate. Trials using these salts in alkyd binders indicate that pigments of this type can provide corrosion protection comparable to that of zinc yellow [39,42,43].

Orthophosphates, the third type of second-generation zinc phosphates, are prepared by reacting orthophosphoric acid with alkaline compounds [12]. This group includes

- Zinc aluminum phosphate. It is formed by combining zinc phosphate and aluminum phosphate in the wet phase; the aluminum ions hydrolyze, causing acidity, which in turn increases the phosphate concentration [12,44,45]. Aluminum phosphate is added to give a higher phosphate content.
- Organically modified basic zinc phosphates. An organic component is fixed onto the surface of basic zinc phosphate particles, apparently to improve compatibility with alkyd and physically drying resins.
- Basic zinc molybdenum phosphate hydrate. Zinc molybdate is added to basic zinc phosphate hydrate so it can be used with water-soluble systems, for example, styrene-modified acrylic dispersions [12]. The pigment produces a molybdate anion (MoO_4^{2-}) that is an effective anodic inhibitor; its passivating capacity is only slightly less than that of the chromate anion [11].

The third generation of zinc phosphates consists of polyphosphates and polyphosphate silicates. Polyphosphates—molecules of more than one phosphorous atom together with oxygen—result from condensation of acid phosphates at higher temperatures than used to produce orthophosphates [12]. This group includes

- Zinc aluminum polyphosphate. This pigment contains a higher percentage of phosphate, as P_2O_5, than zinc phosphate or modified zinc orthophosphates.
- Strontium aluminum polyphosphate. This pigment also has greater phosphate content than first-generation zinc phosphate. The solubility behavior is further altered by inclusion of a metal with an alkaline oxide compared with amphoteric zinc [12].
- Calcium aluminum polyphosphate silicate. This pigment exhibits an altered solubility behavior due to calcium. The composition is interesting: active components are fixed on the surface of an inert filler, wollastonite.
- Zinc calcium strontium polyphosphate silicate. In this pigment, the electrochemically active compounds are also fixed on the surface of wollastonite.

4.2.3   Accelerated Testing and Why Zinc Phosphates Sometimes Fail

Although zinc phosphates show acceptable performance in the field, they commonly show inferior performance in accelerated testing. This response is probably affected by their very low solubility. In accelerated tests, the penetration rate of aggressive ions is highly speeded up, but the solubility of zinc phosphate is not. The amount of aggressive ions thus exceeds the protective capacity of both the phosphate anion and the iron oxide layer on the metal substrate [11]. Bettan has postulated that there is an initial lag time with zinc phosphates because the protective phosphate complex forms slowly on steel's surface. Because the amount of corrosion-initiating ions is increased from the very beginning of an accelerated test, corrosion processes can be initiated during this lag time. In field exposure, lag time is not a problem, because the penetration of aggressive species usually also has its own lag time. Angelmayer has also supported this explanation [40,46].

Romagnoli and Vetere [11] also point out that researcher findings conflict and offers some possible reasons why:

- Experimental variables of the zinc phosphate pigments may differ. One example is distribution of particle diameter; smaller diameter means increased surface area, which increases the amount of phosphate leaching from the pigment. The more phosphate anion in a solution, the better the anticorrosion protection. PVC and CPVC for the particular paint formulations used are also important and frequently neglected. And, of course, because the term *zinc phosphate* applies to both a family of pigments and a specific formula, the exact type of zinc phosphate is important.
- Binder type and additives are not the same. In accelerated testing, the type of binder is usually the most important factor because of its barrier properties. Only after the binder barrier is breached does effect of pigment become apparent.

Good results in accelerated tests have been reported too, though. In a study of maintenance coatings, a zinc phosphate pigmented epoxy primer performed on the same level as zinc epoxies [47].

4.2.4   Aluminum Triphosphate

Hydrated dihydrogen aluminum triphosphate ($AlH_2P_3O_{10} \cdot 2H_2O$) is an acid with a dissociation constant, pKa, of approximately 1.5–1.6. Its acidity per unit mass is approximately 10–100 times higher than that of other similar acids, such as aluminum and silicon hydroxides.

When dissolved, aluminum triphosphate dissociates into triphosphate ions:

$$AlH_2P_3O_{10} \rightarrow Al^{3+} + 2H^+ + \left[P_3O_{10}\right]^{5-}$$

Beland suggests that corrosion protection comes both from the ability of the tripolyphosphate ion to chelate iron ions (passivating the metal) and from tripolyphosphate ions' ability to depolymerize into orthophosphate ions, giving higher phosphate levels than zinc or molybdate phosphate pigments [14].

Chromy and Kaminska attribute the corrosion protection entirely to the triphosphate. They suggest that the anion $(P_3O_{10})^{5-}$ reacts with anodic iron to yield an insoluble layer, which is mainly ferric triphosphate. This phosphate coating is insoluble in water, is very hard, and exhibits excellent adhesion to the substrate [13].

Aluminum triphosphate has limited solubility in water and is frequently modified with either zinc or silicon to control both solubility and reactivity [14,48]. Researchers have demonstrated that aluminum triphosphate is compatible with various binders, including long, medium, and short oil alkyds; epoxies; epoxy-polyesters; and acrylic-melamine resins [49–52]. Chromy and Kaminska note that it is particularly effective on rapidly corroding coatings; it may therefore be useful in overcoating applications [13].

4.2.5 OTHER PHOSPHATES

Phosphate pigments other than zinc and aluminum phosphates have received much less attention in the technical literature. This group includes phosphates, hydroxyphosphates, and acid phosphates of the metals iron, barium, chromium, cadmium, and magnesium. For iron and barium, the only important phosphates appear to be $FePO_4 \cdot 2H_2O$, $Ca_3(PO_4)_2 \cdot \frac{1}{2}H_2O$, $Ba_3(PO_4)_2$, $BaHPO_4$, and $FeNH_4PO_4 \cdot 2H_2O$ [11,13]. Iron phosphate by itself gives poor results, at least in accelerated testing, but appears promising when used with basic zinc phosphate. Reaction accelerators, such as sodium molybdate and sodium m-nitrobenzene sulfonate, have been found to improve the corrosion resistance of coatings containing iron phosphate [53].

Calcium acid phosphate, $CaHPO_4$, has also been discussed in the literature as an anticorrosion pigment. Vetere and Romagnoli have studied it as a replacement for zinc tetroxychromate. When used in a phenolic chlorinated rubber binder, calcium acid phosphate outperformed the simplest zinc phosphate [$Zn_3(PO_4)_2$] and was comparable to zinc tetroxychromate in salt spray testing. However, researchers were not able to identify the mechanism by which this pigment could offer protection to metal. Iron samples in an aqueous suspension of the pigment showed some passivity in corrosion potential measurements. Analysis of the protective layer's composition showed that it is composed mostly of iron oxides; calcium and phosphate ions are present but not, perhaps, at the levels expected for a good passivating pigment [54].

Another phosphate pigment that has been studied is lauryl ammonium phosphate. However, very little information is available about this pigment. Gibson and Camina briefly describes studies using lauryl ammonium phosphate, but the results do not seem to warrant further work with this pigment [16].

4.3 FERRITES

Ferrite pigments have the general formula $MeO \cdot Fe_2O_3$, where Me = Mg, Ca, Sr, Ba, Fe, Zn, or Mn. They are manufactured by calcination of metal oxides. The principal reaction is

$$MeO + Fe_2O_3 \rightarrow MeFe_2O_4$$

at temperatures of approximately 1000°C. These high temperatures translate into high production costs for this class of pigments [14].

Ferrite pigments appear to protect steel both by creating an alkaline environment at the coating–metal interface and, with certain binders, by forming metal soaps. Kresse [44,55] has found that zinc and calcium ferrites react with fatty acids in the binder to form soaps and attributes the corrosion protection to passivation of the metal by the alkaline environment thus created in the coating.

Sekine and Kato [56] agree with this soap formation mechanism. However, they have also tested several ferrite pigments in an epoxy binder, which is not expected to form soaps with metal ions. All the ferrite pigmented epoxy coatings offered better corrosion protection than both the same binder with red iron oxide as anticorrosion pigment and the binder with no anticorrosion pigment. Examination of the rest potential versus immersion time of the coated panels showed a lag time between the initial immersion and passivation of approximately 160 hours in this study. The authors concluded that passivation of the metal occurs only after water has permeated the coating and reached the paint or metal interface [57]. In a study of zinc ferrite in waterborne paints, the ferrite was found to increase the polarization resistance of the steel substrate, that is, passivation. The effect was assumed to be due to formation of an alkaline environment at the steel–coating interface [58].

Sekine and Kato also examined the pH of aqueous extractions of ferrite pigments and the corrosion rate of mild steel immersed in these solutions [57]. Their results are presented in Table 4.3. These data are interesting because they imply that, in addition to soap formation, the pigments can also create an alkali environment at the metal or paint interface. These authors have found that the corrosion-protective properties of the ferrite pigments in epoxy paint films, based on electrochemical

TABLE 4.3
Corrosion Rate of Mild Steel in Extracted Aqueous Solution of Pigments

Pigment	pH	Corrosion Rate, $g/dm^2/day$
Mg ferrite	8.82	12.75
Ca ferrite	12.35	0.26
Sr ferrite	7.85	16.71
Ba ferrite	8.20	18.00
Fe ferrite	8.40	14.95
Zn ferrite	7.31	14.71
Red iron oxide	3.35	20.35
No pigment	6.15	15.82

Source: Sekine, I., and Kato, T., *Ind. Eng. Chem. Prod. Res. Dev.*, 25, 7, 1986. Copyright 1986, American Chemistry Society. Reprinted with permission.

measurements, were (in decreasing order) Mg > Fe > Sr > Ca > Zn > Ba. It should be emphasized that this ranking was obtained in one study: the relative ranking within the ferrite group may owe much to such variables as particle size of the various pigments and PVC (comparable percent weights rather than PVC were used).

Verma and Chakraborty [59] compared zinc ferrite and calcium ferrite to red lead and zinc chromate pigments in aggressive industrial environments. The vehicle used for the pigments was a long oil linseed alkyd resin. Panels were exposed for eight months in five fertilizer plant environments: a urea plant, an ammonium nitrate plant, a nitrogen–phosphorous–potassium (NPK) plant, a sulfuric acid plant, and a nitric acid plant, where, the authors note, acid fumes and fertilizer dust spills are almost continual occurrences. Results vary greatly, depending on plant type. In the sulfuric acid plant, the two ferrites outperformed the lead and chromate pigments by a very wide margin. In the urea and NPK plants, the calcium ferrite pigment was better than any other pigment. In the ammonium nitrate plant, the calcium ferrite pigment performed substantially worse than the others. In the nitric acid plant, the zinc chromate pigment performed significantly worse than the other three, but among these three, the difference was not substantial. The authors attribute the superior behavior of calcium ferrite over zinc ferrite to the former's controlled but higher solubility. Metal ions in solution, they suggest, react with aggressive species that are permeating into the coating and thus prevent them from reaching the metal–coating interface.

4.4 OTHER INHIBITIVE PIGMENTS

Other types of inhibitive pigments include calcium-exchanged silica, barium metaborate, molybdates, and silicates.

4.4.1 CALCIUM-EXCHANGED SILICA

Calcium-exchanged silica is prepared by ion exchanging an anticorrosion cation, calcium, onto the surface of a porous inorganic oxide of silica. The protection mechanism is ion exchange: aggressive cations (e.g., H^+) are preferentially exchanged onto the pigment's matrix as they permeate the coating, while Ca^{2+} ions are simultaneously released to protect the metal. Calcium does not itself passivate the metal or otherwise directly inhibit corrosion. Instead, it acts as a flocculating agent. The small amounts (ca. 120 μm/ml H_2O at pH $\approx$ 9) of silica in solution flocculate around the Ca^{2+} ion. The Ca-Si species has a small $\delta+$ or $\delta-$ charge, which drives it toward the metal surface (due to the potential drop across the metal–solution interface). Particles of silica and calcium agglomerate at the paint–metal interface. There the alkaline pH causes spontaneous coalescing into a thin film of silica and calcium [60]. Auger electron spectroscopy (AES) has confirmed enrichment of silicon on the steel–coating interface after immersion testing [61]. Improved corrosion protection with calcium-exchanged silica has also been attributed to improved barrier properties of the coating [61]. The major benefit of this inorganic film seems to be that it prevents Cl$^-$ and other corrosion-initiating species from reaching the metal surface.

The dual action of entrapment of aggressive cations and release of inhibitor gives calcium-exchanged silica two advantages over traditional anticorrosion pigments:

1. The "inhibitor" ion is only released in the presence of aggressive cations, which means that no excess of the pigment to allow for solubility is necessary.
2. No voids are created in the film by the ion exchange; the coating has fairly constant permeability characteristics [12,60,62,63].

4.4.2 BARIUM METABORATE

Barium metaborate is a pigment to avoid. It contains a high level of soluble barium, an acute toxicant. Due the toxic properties, it has also been used for antimicrobial purposes in paint. Disposal of any waste containing this pigment is likely to be expensive, whether that waste is produced in the manufacture or application of the coatings or much later when preparing to repaint structures originally coated with barium metaborate.

Barium metaborate creates an alkaline environment, inhibiting the steel; the metaborate ion also provides anodic passivation [14]. The pigment requires high loading levels, up to 40% of coating weight, according to Beland. It is highly soluble and fairly reactive with several kinds of binders; this leads to stability problems when formulated with such products as acidic resins, high-acid-number resins, and acid-catalyzed baking systems. A modified silica coating is often used to reduce and control solubility. One way to decrease its reactivity, and therefore increase the number of binders with which it can be used, is to modify it with zinc oxide or a combination of zinc oxide and calcium sulfate [14]. The high loading level required for heavy-duty applications implies that careful attention must be paid to the PVC/CPVC ratio when formulating with this pigment.

Information regarding actual service performance of barium metaborate coatings is scarce, and what does exist does not seem to justify the use of this pigment. In the early 1980s, the state of Massachusetts repair-painted a bridge with barium metaborate pigment in a conventional oil or alkyd vehicle. The result was not satisfactory: after six years, considerable corrosion had occurred at the beam ends and on the railings above the road [1]. It should perhaps be noted that an alkyd vehicle is not the ideal choice for a pigment that generates an alkaline environment; better results may perhaps have been obtained with a higher-performance binder. However, because of the toxicity problems associated with soluble barium, further work with barium metaborate does not seem to be warranted.

4.4.3 MOLYBDATES

Molybdate pigments are calcium or zinc salts precipitated onto an inert core such as calcium carbonate [22,64–66]. They prevent corrosion by inhibiting the anodic corrosion reaction [22]. The protective layer of ferric molybdate, which these pigments form on the surface of the steel, is insoluble in neutral and basic solutions.

Use of these pigments has been limited because of their expense. Zinc phosphate versions of the molybdate pigments have been introduced in order to lower costs and improve both adhesion to steel and film flexibility [14,22,64–66]. The molybdate pigment family includes

- Basic zinc molybdate
- Basic calcium zinc molybdate
- Basic zinc molybdate
- Basic calcium zinc molybdate

In general, tests of these pigments as corrosion inhibitors in paint formulations have returned mixed results on steel. Workers in the field tend to refer somewhat wistfully to the possibilities of improving the performance of molybdates through combination with other pigments, in the hope of obtaining a synergistic effect. A serious drawback is that, in several studies, molybdates appeared to cause coating embrittlement, perhaps due to premature binder aging [67–70].

Although molybdate pigments are considered nontoxic [71], they are not completely harmless. When cutting or welding molybdate pigmented coatings, fumes of very low toxicity are produced. With proper ventilation, these fumes are not likely to prove hazardous [68]. The possible toxicity is about 10%–20% that of chromium compounds [71,72].

4.4.4 Silicates

Silicate pigments consist of soluble metallic salts of borosilicates and phosphosilicates. The metals used in silicate pigments are barium, calcium, strontium, and zinc; silicates containing barium can be assumed to pose toxicity problems.

The silicate pigments include

- Calcium borosilicates, which are available in several grades, with varying B_2O_3 content (not suitable for immersion or semi-immersion service or water-based resins [14])
- Calcium barium phosphosilicate
- Calcium strontium phosphosilicate
- Calcium strontium zinc phosphosilicate, which is the most versatile phosphosilicate inhibitor in terms of binder compatibility [14]

The silicate pigments can inhibit corrosion in two ways: through their alkalinity and, in oleoresinous binders, by forming metal soaps with certain components of the vehicle. Which process predominates is not entirely clear, perhaps because the efficacy of the pigments is not entirely clear. When Heyes and Mayne examined calcium phosphosilicate and calcium borosilicate pigments in drying oils, they found a mechanism similar to that of red lead: the pigment and the oil binder react to form metal soaps, which degrade and yield products with soluble, inhibitive anions [73].

Van Ooij and Groot found that calcium borosilicate worked well in a polyester binder, but not in an epoxy or polyurethane [74]. This hints that the alkalinity generated within the binder cannot be very high; otherwise, the polyester—being much more vulnerable to saponification—would have shown much worse results than either the epoxy or the polyurethane. Metal soaps, of course, would not be formed with either an epoxy or polyurethane. However, the possibility of metal soaps cannot be absolutely ruled out for a polyester without knowing exactly what is meant by this unfortunately broad term.

The state of Massachusetts had a less positive experience with the same pigment, although possibly a different grade of it. In the 1980s, Massachusetts repair-painted a number of bridges with calcium borosilicate pigment in a conventional oleoresinous binder—a vehicle that would presumably form metal soaps. Spot blasting was performed prior to coating. The calcium borosilicate system was judged less forgiving of poor surface preparation than lead-based paint (LBP), and attaining the minimum film build was found to be critical. Massachusetts eventually stopped using this pigment because of the high costs of improved surface preparation and inspection of film build.

Another silicate, calcium barium phosphosilicate, has been tested in conjunction with six other pigments on cold-rolled steel in an epoxy-polyamide binder. After nine months' atmospheric exposure in a marine environment (Biarritz, France), the samples with calcium barium phosphosilicate pigment—and those with barium metaborate—gave worse results than either the aluminum triphosphate or ion-exchanged calcium silicate pigments. (These in turn were significantly outperformed by a modified zinc phosphate, as well as by a zinc chromate pigment.)

4.5 BARRIER PIGMENTS

Barrier coatings protect steel against corrosion by reducing the permeability of liquids and gases through a paint film. How much the permeability of water and oxygen can be reduced depends on many factors, including

- Thickness of the film
- Structure of the film (polymer type used as binder)
- Degree of binder crosslinking
- PVCs
- Type and particle shape of pigments and fillers

Pigments used for barrier coatings are diametrically opposed to the active pigments used in other anticorrosion coatings in one respect: in barrier coatings, they must be inert and completely insoluble in water. Commonly used barrier pigments can be broken into two groups:

- Mineral-based materials, such as mica, micaceous iron oxide (MIO), and glass flakes
- Metallic flakes of aluminum, zinc, stainless steel, nickel, and cupronickel

In the second group, care must be taken to avoid possible electrochemical interactions between the metallic pigments and the metal substrate [75].

Flake-shaped pigments tend to orient themselves parallel to the substrate, as can be seen in Figures 1.1 and 3.3. This is probably due to lateral flow of the wet paint during application. In spray painting, droplets of paint hit the surface and flow out into a film, which means that the paint will flow parallel to the substrate. The flake-shaped pigments will then mainly be oriented parallel with the flow, since this creates the least resistance against the flow. Surface treatment of the pigments and application parameters will affect the orientation [76].

4.5.1 MICACEOUS IRON OXIDE

MIO is a naturally occurring iron oxide pigment that contains at least 85% Fe_2O_3. The term *micaceous* refers to its particle shape, which is flake-like or lamellar: particles are very thin compared with their area. This particle shape is extremely important for MIO in protecting steel. MIO particles orient themselves within the coating, so that the flakes are lying parallel to the substrate's surface. Multiple layers of flakes form an effective barrier against moisture and gases [15,75,77–83]. MIO is fascinating in one respect: it is a form of rust that has been used as an effective pigment in barrier coatings for decades to protect steel from rusting.

For effective barrier properties, PVCs in the range of 25%–45% are used, and the purity must be at least 80% MIO (by weight). Because MIO is a naturally occurring mineral, it can vary from source to source, both in chemical composition and in particle size distribution. Smaller flakes mean more layers of pigment in the dried film, which increases the pathway that water must travel to reach the metal. Schmid has noted that in a typical particle size distribution, as much as 10% of the particles may be too large to be effective in thin coatings, because there are not enough layers of flakes to provide a barrier against water. To provide a good barrier in the vicinity of these large particles, MIO is used in thick coatings or multicoats [84].

Historically, it has been believed that MIO coatings tend to fail at sharp edges because the miox particles were randomly oriented in the vicinity of edges. Random orientation would, of course, increase the capillary flow of water along the pigment's surface toward the metal substrate. However, Wiktorek and Bradley examined coverage over sharp edges using scanning electron microscope images of cross sections. They found that lamellar miox particles always lie parallel to the substrate, even over sharp edges. The authors suggested that when failure is seen at edges, the problem is really thinner coatings in these areas [85].

In addition to providing a barrier against diffusion of aggressive species through the coating, MIO confers other advantages:

- It provides mechanical reinforcement to the paint film.
- It can block UV light, thus shielding the binder from this destructive form of radiation.

For the latter reason, MIO is sometimes used in topcoat formulations to improve weatherability [15,75].

The chemical inertness of MIO means that it can be used in a variety of binders: alkyd, chlorinated rubber, styrene-acrylic and vinyl copolymers, epoxy, and polyurethane [15].

It is not clear from the literature whether combining MIO and aluminum pigments in a coating poses a problem. There are recommendations both for and against mixing MIO with these pigments.

In full-scale trials of various paint systems on bridges in England, Bishop found that topcoats with both MIO and aluminum pigments form a white deposit over large areas. Analysis showed these deposits to be mostly aluminum sulfate with some ammonium sulfate. The only possible source of aluminum in the coating system was the topcoat pigment. Bishop did not find the specific cause of this problem. He notes that bridge paints in the United States commonly contain leafing aluminum and that few problems are reported [86].

Schmid, on the other hand, recommends combining MIO with other lamellar materials, such as aluminum flake and talc, to improve the barrier properties of the film by closer pigment packing [84].

4.5.2 MICA

Mica is a group of hydrous potassium aluminosilicates. The diameter-to-thickness ratio of this group exceeds 25:1, higher than that of any other flaky pigment. This makes mica very effective at building up layers of pigment in the dried film, thus increasing the pathway that water must travel to reach the metal and reducing water permeability [87,88].

4.5.3 GLASS

Glass fillers include flakes, beads, microspheres, fibers, and powder. Glass flakes provide the best coating barrier properties. Other glass fillers can also form a protective barrier because of their close packing in the paint coating. Glass has been used in the United States, Japan, and Europe when high-temperature resistance, or high resistance to abrasion, erosion, and impact, is needed. The thicknesses of coatings filled with glass flakes are approximately 1–3 mm; flakes are 3–5 µm thick, so every millimeter of coating can contain approximately 100 layers of flakes [75].

Studies have shown that glass flakes perform comparably to lamellar pigments of stainless steel and MIO pigments, but perform worse than aluminum flake; the latter showed better flake orientation than glass flake in the paint film [75,89–91]. Glass flake is usually preferred for elevated temperatures, not only because of its ability to maintain chemical resistance at high temperatures but also because of its coefficient of thermal expansion. Coatings filled with glass flake can obtain thermal expansion properties close to those of carbon steel. This enables them to retain good adhesion even under thermal shock [92,93].

Glass beads, microspheres, fibers, and powders are also used for their thermal properties in fire-resistant coatings. Spherical glass beads can increase the mechanical strength of a cured film. Using beads of various diameters can improve packing inside the dry film, thus improving barrier properties. Glass fibers impart good abrasion resistance to the paint. Glass microspheres are a component of the fly ash produced by the electric power industry. More precisely, they are aluminosilicate spheres, with diameters between 0.3 and 200 μm, that are composed of Al_2O_3, Fe_2O_3, CaO, MgO, Na_2O, and K_2O. The exact makeup depends on the type and source of fuel burned [75].

4.5.4 ALUMINUM

Besides reducing the permeability of water vapor, oxygen, and other corrosive media, aluminum pigment also reflects UV radiation and can withstand elevated temperatures. There are two types of aluminum pigment: leafing and nonleafing. Leafing pigment orients itself parallel to the substrate at the top of the coating; this positioning enables the pigment to protect the binder against UV damage but may not be the best location for maximizing barrier properties. Leafing properties depend on the presence of a thin fatty acid layer, commonly stearic acid, on the flakes. Nonleafing aluminum pigments are distributed inside the coating film. They tend to orient themselves parallel with the substrate and are very effective in barrier coatings [75]. Aluminum flake pigments are also used in powder coatings and waterborne paint. In waterborne paint, the aluminum flakes must be surface treated to protect them from corrosion under hydrogen evolution inside the container [94]. Aluminum flakes have been shown to have a beneficial effect on cathodic disbonding [95,96]. The effect was attributed to a buffering effect, where the aluminum pigments corroded inside the paint, consuming the high-pH environment at the steel–coating interface that causes cathodic disbonding (Section 12.1), and not a barrier effect.

4.5.5 ZINC FLAKES

Zinc flakes should not to be confused with the zinc dust used in zinc-rich coatings: the size is of a different magnitude altogether. Some research suggests that zinc flakes could give both the cathodic protection typical of zinc dust and the barrier protection characteristic of lamellar pigments [75]. However, in practice, this could be very difficult to achieve because the zinc dust particles in ZRPs have to be in electrical contact to obtain cathodic protection. Designing a coating in which the zinc particles are in intimate contact with each other and with the steel, and yet completely free of gaps between pigment and binder or between pigment particles, is difficult. The lack of any gaps is critical for a barrier pigment, because it is precisely these gaps that provide the easy route for water and oxygen to reach the metal surface. In fact, Hare and Wright's [97] research shows that zinc flakes undergo rapid dissolution in corrosive environments when they are used as the sole pigment in paints; their coatings are prone to blistering.

4.5.6 OTHER METALLIC PIGMENTS

Other metallic pigments, such as stainless steel, nickel, and copper, have also been used. Their use in coatings of metals with more noble electrochemical potential than carbon steel entails a certain risk of galvanic corrosion between the coating and the substrate. The PVCs in such paints must be kept well below the levels at which the metallic pigment particles are in electrical contact with each other and the carbon steel. If this condition is not met, accelerated coating degradation and corrosion follow. Bieganska recommends using a nonconducting primer as an insulating layer between the steel substrate and the barrier coating, if it is necessary to use a noble and conducting pigment in the barrier layer [75]. The same author also warns that although the mechanical durability and high-temperature resistance of stainless steel flake make this type of pigment desirable, it is not suited to applications where chlorides are present [75].

Nickel flake-filled coatings can be useful for strongly alkaline environments. Cupronickel flakes (Cu–10% Ni–2% Sn) are used in ship protection because of their outstanding antifouling properties. The alloy pigment is of interest in this application because its resistance to leaching is better than that of copper itself [75].

4.6   CHOOSING A PIGMENT

Before choosing a pigment and formulating paint, one question must be answered: Will an active or a passive role be required of the pigment? The role of the pigment—active or passive—must be decided at the start for the fairly straightforward reason that only one or the other is possible. Many of the pigments that actively inhibit corrosion, such as through passivation, must dissolve into anions and cations; ion species can then passivate the metal surface. Without water, these pigments do not dissolve and the protection mechanism is not triggered. And, of course, it is the express purpose of barrier coatings to prevent water from reaching the coating–metal interface.

Once the role of the pigment has been decided, choice of pigment depends on such factors as

- *Price.* Many of the newer pigments are expensive. The amounts necessary in a coating, and the respective impact on price, play a large role in determining whether the pigment is economically feasible.
- *Commercial availability.* Producing a few hundred grams of a pigment in a laboratory is one thing; however, it is quite another to generate a pigment in hundreds of kilograms for commercial paints.
- *Difficulty of blending into a real formulation.* Pigments must do more than just protect steel. They have to disperse in the wet paint, rather than stay clumped together. They also have to be well attached to the binder so that water cannot penetrate through the coating via gaps between pigment particles and the binder. In many cases, the surfaces of pigments are chemically treated to avoid these problems; however, it must be possible to treat pigments without changing their essential properties (solubility, etc.).
- *Suitability in the binders that are of interest.* A coating does not, of course, consist merely of a pigment; the binder is of equal importance in determining the success of a paint. The pigments chosen for further study must be compatible with the binders of interest.
- Resistance to heat, acids or alkalis, or solvents, as needed.

4.7   ABANDONED PIGMENTS DUE TO TOXICITY

The toxicity of lead, chromium, cadmium, and barium has made the continued use
of paints containing these elements highly undesirable. The health and environmental problems associated with these heavy metals are serious, and new problems are
continuously discovered. To address this issue, pigment manufacturers have developed many alternative pigments, like the ones described in the previous sections.
The number of proposed alternatives is not lacking; in fact, the number and types
available are nearly overwhelming.

The toxic pigments listed above are almost completely abandoned in coatings, but
lead and chromate are included in this section because the protective mechanism is
interesting when developing new pigments.

4.7.1   LEAD-BASED PAINT

The inhibitive mechanism of the red lead found in LBP is complex. Lead pigments
may be thought of as indirect inhibitors because although they themselves are not
inhibitive, they undergo a reaction with select resin systems, and this reaction can
form by-products that are active inhibitors [14].

Appleby and Mayne [98,99] have shown that formation of lead soaps is the mechanism used for protecting clean (or new) steel. When formulated with linseed oil,
lead reacts with components of the oil to form soaps in the dry film; these soaps
degrade to, among other things, the water-soluble salt lead of a variety of mono-
and dibasic aliphatic acids [100,101]. Mayne and Van Rooyen also showed that the
lead salts of azelaic, suberic, and pelargonic acid were inhibitors of iron corrosion.
Appleby and Mayne have suggested that these acids inhibit corrosion by bringing
about the formation of insoluble ferric salts, which reinforce the air-formed oxide
film until it becomes impermeable to ferrous ions. This finding was based on experiments in which pure iron was immersed in a lead azelate solution, with the thickness
of the oxide film measured before and after immersion. They found that the oxide
film increased 7%–17% in thickness upon immersion [98,102].

The lead salt of azelaic acid dissociates in water into a lead ion and an azelate
ion. To determine which element was the key in corrosion inhibition, Appleby
and Mayne also repeated the experiment with calcium azelate and sodium azelate
[99,103]. Interestingly, they did not see a similar thickening of the oxide film when
iron was immersed in calcium azelate and sodium azelate solutions, demonstrating
that lead itself—not just the organic acid—plays a role in protecting the iron. The
authors note that 5–20 ppm lead azelate in water is enough to prevent attack of pure
iron immersed in the solution. They noted that at this low concentration, inhibition
cannot be caused by the repair of the air-formed oxide film by the formation of a
complex azelate, as is the case in more concentrated solutions; rather, it appears to
be associated with the thickening of the air-formed oxide film.

It seems possible that, initially, lead ions in solution may provide an alternative
cathodic reaction to oxygen reduction, and then, on being reduced to metallic lead at
the cathodic areas on the iron surface, depolarize the oxygen reduction reaction, thus

keeping the current density sufficiently high to maintain ferric film formation. In addition any hydrogen peroxide so produced may assist in keeping the iron ions in the oxide film in the ferric condition, so that thickening of the air-formed film takes place until it becomes impervious to iron ions. [98]

Protecting rusted steel, rather than clean or new steel, may demand of a paint a different corrosion mechanism, simply because the paint is not applied directly to the steel that must be protected, but rather to the rust on top of it. Inhibitive pigments in the paint that require intimate contact with the metallic surface in order to protect it may therefore not perform well when a layer of rust prevents that immediate contact. Red lead paint, however, does perform well on rusted steel. Several theories about the protective mechanism of red lead paint on rusted steel exist.

A rust impregnation theory has been proposed. The low viscosity of the vehicle used in LBP allows it to penetrate the surface texture of rust. This would have several advantages:

- Impregnation of the rust means that it is isolated and thereby inhibited in its corroding effect.
- Oil-based penetrants provide a barrier effect, thus screening the rust from water and oxygen and slowing down corrosion [48].
- Good penetration and wetting of the rust by the paint results in better adhesion.

Thomas examined cross sections of LBP and other paints on rusted steel using transmission electron microscopy [104,105]; she found that although the paint penetrated well into cracks in the rust layer, there was no evidence that the LBP penetrated through the compact rust layers to the rust–metal interface. (It should be noted that this experiment used cooked linseed oil, not raw; Thomas notes that raw linseed oil has a lower viscosity and might have penetrated further.) Where lead was found, it was always in the vicinity of the paint–rust interface, and in low concentrations. It had presumably diffused into the rust layer after dissolution or breakdown of the red lead pigment and was not present as discrete particles of Pb_3O_4. Thomas also found that the penetration of LBP into the rust layer was not significantly better than that of the other vehicles studied, for example, epoxy mastic. Finally, the penetration rate of water through linseed oil–based LBP was found to be approximately 214 g/m^2/day for a 25-μm film, and that of oxygen was 734 cc/m^2/day for a 100-μm film [105]. The amounts of water and oxygen available through the paint film are greater than the minimum needed for the corrosion of uncoated steel. Therefore, barrier properties can be safely eliminated as the protective mechanism. Superior penetration and wetting do not appear to be the mechanisms by which LBP protects rusted steel.

LBP may protect rusty steel by insolubilizing sulfate and chloride, rendering these aggressive ions inert. Soluble ferrous salts are converted into stable, insoluble, and harmless compounds; for example, sulfate nests can be rendered "harmless' by treatment with barium salts because barium sulfate is extremely insoluble. This was suggested as a protective mechanism of LBP by Lincke and Mahn [106] because, when red lead pigmented films were soaked in concentrated solutions of

Fe(II) sulfate, Fe(III) sulfate, and Fe(III) chloride, precipitation reactions occurred. Thomas [107,108] tested this theory by examining cross sections of LBP on rusted steel (after three years' exposure of the coated samples) using laser microprobe mass spectrometry (LAMMS) and transmission electron microscopy with energy-dispersive x-ray. Low levels of lead were found in the rust layer, but only within 30 μm of the rust–paint interface. Lead was not seen at or near the rust–metal interface, where sulfate nests are known to exist, nor was it distributed throughout the rust layer, even though sulfur was. If rendering inert is truly the mechanism, $PbSO_4$ would be formed as the insoluble "precipitate" within the film, and the ratio of Pb to S would be 1.0 or greater (assuming a surplus of lead exists). However, no correlation was seen between the distribution of lead and that of sulfur (confirmed as sulfate by x-ray photoelectron spectroscopy); the ratio of lead to sulfur was 0.2–1.0, which Thomas concludes is insufficient to protect the steel. Sulfate insolubilization does not, therefore, seem to be the mechanism by which LBP protects rusted steel.

In the previously described work, low levels of lead were found in the rust layer near the paint–rust interface, within 30 μm of the interface. Thomas suggests that because lead salts do not appear to reach the metal substrate to inhibit the anodic reaction, it is possible that lead acts within the rust layer to slow down atmospheric corrosion by interfering with the cathodic reaction (i.e., by inhibiting the cathodic reduction of existing rust [principally FeOOH to magnetite]) [108]. This presumably would suppress the anodic dissolution of iron because that reaction ought to be balanced by the cathodic reaction. No conclusive proof for or against this theory has been offered.

Finally, the lead soap–lead azelate theory has also been proposed to act on rusty steel, like on new steel. Thomas looked for lead (as a constituent of lead azelate) at the steel–rust interface in an attempt to confirm this theory. Samples coated with LBP were exposed for three years, and then cross sections were examined in a LAMMS; however, lead was not detected at the interface. As Thomas points out, this finding does not eliminate the mechanism as a possibility; lead could still be present but in levels below the 100 ppm detection limit of the LAMMS [104,105]. Appleby and Mayne have shown that 5–20 ppm of lead azelate is enough to protect pure iron [98]. The levels needed to protect rusted steel would not be expected to be so low, because the critical concentration required for anodic inhibitors is higher when chloride or sulfate ions are present than when used on new or clean steel [109]. Possibly, a level between 20 and 100 ppm of lead azelate is sufficient to protect the steel. Another point worth considering is that the amounts of lead that would exist in the passive film formed by complex azelates, suggested by Appleby and Mayne, has not been determined. The lead soap–lead azelate theory appears to be the most likely mechanism to explain how red lead paints protect rusted steel.

In summary, formation of lead soaps appears to be the mechanism by which LBPs inhibit corrosion of clean steel, and possibly also on rusty steel. When formulated with linseed oil, lead reacts with components of the oil to form soaps in the cured film; in the presence of water and oxygen, these soaps degrade to, among other things, salts of a variety of mono- and dibasic aliphatic acids. The lead salts of azelaic, suberic, and pelargonic acid act as corrosion inhibitors; lead azelate is of particular importance in LBP. These acids may inhibit corrosion by bringing about the formation of insoluble ferric salts, which reinforce the metal's oxide film until it becomes impermeable to ferrous ions, thus suppressing the corrosion mechanism. The formation of lead soaps is believed to be the critical corrosion protection step for both new (clean) steel and rusted steel.

4.7.2 CHROMATES

The chromate passivating ion is among the most efficient passivators known. However, due to health and environmental concerns associated with hexavalent chromium, this class of anticorrosion pigments is rapidly disappearing.

Simply put, chromate pigments stimulate the formation of passive layers on metal surfaces [110]. The actual mechanism is probably more complex. Svoboda and Mleziva have described the protection mechanism of chromates as "a process which begins with physical adsorption which is transformed to chemisorption and leads to the formation of compounds which also contain trivalent chromium" [111].

In the mechanism described by Rosenfeld et al. [112], CrO_4^{2-} groups are adsorbed onto the steel surface, where they are reduced to trivalent ions. These trivalent ions participate in the formation of the complex compound $FeCr_2O_{14-n}(OH^-)_n$, which in turn forms a protective film. Largin and Rosenfeld have proposed that chromates do not merely form a mixed oxide film at the metal surface; instead, they cause a change in the structure of the existing oxide film, accompanied by a considerable increase in the bond energy between the iron and oxygen atoms. This leads to an increase in the protective properties of the film [113].

It should perhaps also be noted that several workers in the field describe the protection mechanism more simply as the formation of a normal protective mixed oxide film, with defects in the film plugged by Cr_2O_3 [14,32].

The principal chromate-based pigments are basic zinc potassium chromate (also known as zinc yellow or zinc chrome), strontium chromate, and zinc tetroxychromate. Other chromate pigments exist, such as barium chromate, barium potassium chromate, basic magnesium chromate, calcium chromate, and ammonium dichromate; however, because they are used to a much lesser extent, they are not discussed here.

Zinc potassium chromate is the product of inhibitive reactions among potassium dichromate, zinc oxide, and sulfuric acid. Zinc chromates are effective inhibitors even at relatively low loading levels [14].

Strontium chromate, the most expensive chromate inhibitor, is mainly used on aluminum. It is used in the aviation and coil coating industries because of its effectiveness at very low loadings.

Zinc tetroxychromate, or basic zinc chromate, is commonly used in the manufacture of two-package polyvinyl butyryl (PVB) wash primers. These consist of phosphoric acid and zinc tetroxychromate dispersed in a solution of PVB in alcohol. These etch primers, as they are known, are used to passivate steel, galvanized steel, and aluminum surfaces, improving the adhesion of subsequent coatings. They tend to be low in solids and are applied at fairly low film thicknesses [14].

The ability of a chromate pigment to protect a metal lies in its ability to dissolve and release chromate ions. Controlling the solubility of the pigment is critical for chromates. If the solubility is too high, other coating properties, such as blister formation, are adversely affected. A coating that uses a highly soluble chromate pigment under long-term moisture conditions can act as a semipermeable membrane: with water on one side (at the top of the coating) and a saturated solution of aqueous pigment extract on the other (at the steel–coating interface). Significant osmotic forces thus lead to blister formation [111]. Chromate pigments are therefore not suitable for use in immersion conditions or conditions with long periods of condensation or other moisture exposure.

REFERENCES

 1. Hare, C.H. *Mod. Paint Coat.* 76, 38, 1986.
 2. Boxall, J. *Polym. Paint Colour J.* 181, 443, 1991.
 3. Undrum, H. *J. Prot. Coat. Linings* 23, 52, 2006.
 4. Ault, J.P. *J. Prot. Coat. Linings* 28, 11, 2011.
 5. Felui, S., Barajas, R., Bastidas, J., and M. Morcillo. *J. Coating. Tech.* 61, 775, 71–76, 1989.
 6. Zimmerman, K. *Eur. Coat. J.* 1, 14, 1991.
 7. ISO 12944-5. Paints and varnishes—Corrosion protection of steel structures by protective paint systems. Part 5: Protective paint systems. Geneva: International Organization for Standardization, 2007.
 8. Fultz, B.S. Zinc rich coatings—When to topcoat and when not to. Presented at CORROSION/06. Houston: NACE International, 2006, paper 06005.
 9. Zhang, X.G. *Corrosion and Electrochemistry of Zinc.* New York: Springer, 1996.
10. Mitchell, M. Corrosion under insulation. New approaches to coating & insulation materials. Presented at CORROSION/2003. Houston: NACE International, 2003, paper 3036.
11. Romagnoli, R., and V.F. Vetere. *Corros. Rev.* 13, 45, 1995.
12. Krieg, S. *Pitture Vernici* 72, 18, 1996.
13. Chromy, L. A., and E. Kaminska. *Prog. Org. Coat.* 18, 319, 1990.
14. Beland, M. *Am. Paint Coat. J.* 6, 43, 1991.
15. Boxall, J. *Polym. Paint Colour J.* 179, 129, 1989.
16. Gibson, M.C., and M. Camina. *Polym. Paint Colour J.* 178, 232, 1988.
17. Ruf, J. *Werkst. Korros.* 20, 861, 1969.
18. Meyer, G. *Farbe Lack* 68, 315, 1962.
19. Meyer, G. *Farbe Lack* 1963.
20. Meyer, G. *Werkst. Korros.* 16, 508, 1963.
21. Meyer, G. *Farbe Lack* 71, 113, 1965.
22. Boxall, J. *Paint Resin* 55, 38, 1985.
23. Ginsburg, T. *J. Coat. Technol.* 53, 23, 1981.
24. Gooma, A.Z., and H.A. Gad. *J. Oil Colour Chem. Assoc.* 71, 50, 1988.
25. Svoboda, M. *Farbe Lack* 92, 701, 1986.
26. Robu, C., et al. *Polym. Paint Colour J.* 177, 566, 1987.
27. Bernhard, A., et al. *Eur. Suppl. Polym. Paint Colour J.* 171, 62, 1981.
28. Ruf, J. *Chimia* 27, 496, 1973.
29. Dean, S.W., et al. *Mater. Perform.* 12, 47, 1981.
30. Kwiatkowski, L., et al. *Powloki Ochr.* 14, 89, 1988.
31. Pryor, M.J., and M. Cohen. *J. Electrochem. Soc.* 100, 203, 1953.
32. Leidheiser, H., Jr. *J. Coat. Technol.* 53, 29, 1981.
33. Kozlowski, W., and J. Flis. *Corros. Sci.* 32, 861, 1991.
34. Kaminski, W. *J. Prot. Coat. Linings* 13, 57, 1996.

35. Clay, M.F., and J.H. Cox. *J. Oil Colour Chem. Assoc.* 56, 13, 1973.
36. Szklarska-Smialowska, Z., and J. Mankowsky. *Br. Corros. J* 4, 271, 1969.
37. Burkill, J.A., and J.E.O. Mayne. *J. Oil Colour Chem. Assoc.* 9, 273, 1988.
38. Bittner, A. *J. Coat. Technol.* 61, 111, 1989.
39. Adrian, G. *Polym. Paint Colour J.* 175, 127, 1985.
40. Bettan, B. *Paint Resin* 56, 16, 1986.
41. Bettan, B. *Pitture Vernici* 63, 33, 1987.
42. Adrian, G., et al. *Farbe Lack* 87, 833, 1981.
43. Bittner, A. *Pitture Vernici* 64, 23, 1988.
44. Kresse, P. *Farbe Lack* 83, 85, 1977.
45. Gerhard, A., and A. Bittner. *J. Coat. Technol.* 58, 59, 1986.
46. Angelmayer, K.-H. *Polym. Paint Colour J.* 176, 233, 1986.
47. Bjørgum, A., et al. Repair coating systems for bare steel: Effect of pre-treatment and conditions during application and curing. Presented at CORROSION/2007. Houston: NACE International, 2007, paper 07012.
48. Van Oeteren, K.A. *c* 73, 12, 1971.
49. Nakano, J. *Polym. Paint Colour J.* 175, 328, 1985.
50. Nakano, J. *Polym. Paint Colour J.* 175, 704, 1985.
51. Nakano, J. *Polym. Paint Colour J.* 177, 642, 1987.
52. Takahashi, M. *Polym. Paint Colour J.* 177, 554, 1987.
53. Gorecki, G. *Metal Finish.* 90, 27, 1992.
54. Vetere, V.F., and R. Romagnoli. *Br. Corros. J.* 29, 115, 1994.
55. Kresse, P. *Farbe Lack* 84, 156, 1978.
56. Sekine, I., and T. Kato. *J. Oil Colour Chem. Assoc.* 70, 58, 1987.
57. Sekine, I., and T. Kato. *Ind. Eng. Chem. Prod. Res. Dev.* 28, 7, 1986.
58. Zubielewicz, M., and W. Gnot. *Prog. Org. Coat.* 49, 358, 2004.
59. Verma, K.M., and B.R. Chakraborty. *Anti-Corros. Methods Mater.* 34, 4, 1987.
60. Goldie, B.P.F., *J. Oil Colour Chem. Assoc.* 71, 257, 1988.
61. Vasconcelos, L.W., et al. *Corros. Sci.* 43, 2291, 2001.
62. Goldie, B.P.F. *Paint Resin* 1, 16, 1985.
63. Goldie, B.P.F. *Polym. Paint Colour J.* 175, 337, 1985.
64. Banke, W.J. *Mod. Paint Coat.* 2, 45, 1980.
65. Garnaud, M.H.L. *Polym. Paint Colour J.* 174, 268, 1984.
66. Sullivan, F.J., and M.S. Vukasovich. *Mod. Paint Coat.* 3, 41, 1981.
67. Lapasin, R., et al. *J. Oil Colour Chem. Assoc.* 58, 286, 1975.
68. Lapasin, R., et al. *Br. Corros. J.* 12, 92, 1977.
69. Marchese, A., et al. *Anti-Corros. Methods Mater.* 23, 4, 1974.
70. Wilcox, G.D., et al. *Corros. Rev.* 6, 327, 1986.
71. Sherwin-Williams Chemicals. Technical Bulletin 342. New York: Sherwin-Williams Chemicals.
72. ACGIH (American Conference of Governmental Industrial Hygienists). *Threshold Limit Values for Chemical Substances and Biological Exposure Indices.* Vol. 3, Report 1971. Cincinnati, OH: ACGIH.
73. Heyes, P.J., and J.E.O. Mayne. Inhibitive pigments for protection of iron and steel in *6th European Congress on Metallic Corrosion*, London, 1977, paper 213.
74. Van Ooij, W.J., and R.C. Groot. *J. Oil Colour Chem. Assoc.* 69, 62, 1986.
75. Bieganska, B., et al. *Prog. Org. Coat.* 16, 219, 1988.
76. Maile, F.J., et al. *Prog. Org. Coat.* 54, 150, 2005.
77. Bishop, D.M. *J. Oil Colour Chem. Assoc.* 66, 67, 1981.
78. Bishop, D.M., and F.G. Zobel. *J. Oil Colour Chem. Assoc.* 66, 67, 1983.
79. Boxall, J. *Polym. Paint Colour J.* 174, 272, 1984.
80. Carter, E. *Polym. Paint Colour J.* 171, 506, 1981.

81. Schmid, E.V. *Farbe Lack* 90, 759, 1984.
82. Schuler, D. *Farbe Lack* 92, 703, 1986.
83. Wiktorek, S., and J. John. *J. Oil Colour Chem. Assoc*. 66, 164, 1983.
84. Schmid, E.V. *Polym. Paint Colour J*. 181, 302, 1991.
85. Wiktorek, S., and E.G. Bradley. *J. Oil Colour Chem. Assoc*. 69, 172, 1986.
86. Bishop, R.R. *Mod. Paint Coat*. 9, 149, 1974.
87. Oil and Colour Chemists' Association. *Surface Coatings*. Vol. 1. London: Chapman & Hall, 1983.
88. Eickhoff, A.J. *Mod. Paint Coat*. 67, 37, 1977.
89. El-Sawy, S.M., and N.A. Ghanem. *J. Oil Colour Chem. Assoc*. 67, 253, 1984.
90. Hare, C.H. *Mod. Paint Coat*. 75, 37, 1985.
91. Hare, C.H., and M.G. Fernald. *Mod. Paint Coat*. 74, 138, 1984.
92. Hearn, R.C. *Corros. Prev. Control* 34, 10, 1987.
93. Sprecher, N. *J. Oil Colour Chem. Assoc*. 66, 52, 1983.
94. Karlsson, P., Palmqvist, A.E.C., and K. Holmberg, Surface modification for aluminium pigment inhibition, *Adv. Colloid Interface Sci.,* 128–130, 121–134, 2006.
95. Knudsen, O.Ø., and U. Steinsmo. *J. Corros. Sci. Eng.* 2, 13, 1999.
96. Knudsen, O.Ø., and U. Steinsmo. *J. Corros. Sci. Eng.* 2, 37, 1999.
97. Hare, C.H., and S.J. Wright. *J. Coat. Technol.* 54, 65, 1982.
98. Appleby, A.J., and J.E.O. Mayne. *J. Oil Colour Chem. Assoc*. 50, 897, 1967.
99. Appleby, A.J., and J.E.O. Mayne. *J. Oil Colour Chem. Assoc*. 59, 69, 1976.
100. Mayne, J.E.O., and E.H. Ramshaw. *J. Appl. Chem.* 13, 553, 1963.
101. Mayne, J.E.O., and D. Van Rooyden. *J. Appl. Chem.* 4, 419, 1960.
102. Hancock, P. *Chem. Ind.* 194, 1961.
103. Rychla, L. *Int. J. Polym. Mater.* 13, 227, 1990.
104. Thomas, N.L. The protective action of red lead pigmented alkyds on rusted mild steel, in *Proceedings of the* Symposium on Advances in Corrosion Protection by Organic Coatings. Pennington, NJ: Electrochemical Society, The protective action of red lead pigmented alkyds on rusted mild steel 1989, 451.
105. Thomas, N.L. *Prog. Org. Coat.* 19, 101, 1991.
106. Lincke, G., and W.D. Mahn. In *Proceedings of the 12th FATIPEC Congress*. Paris: Fédération d'Associations de Techniciens des Industries des Peintures, Vernis, Emaux et Encres d'Imprimerie de l'Europe Continentale (FATIPEC), 1974, 563.
107. Thomas, N.L. *J. Prot. Coat. Linings* 6, 63, 1989.
108. Thomas, N.L. Coatings for difficult surfaces. Presented at Proceedings of PRA Symposium, Hampton, UK, 1990, paper 10.
109. Brasher, D., and A. Kingsbury. *J. Appl. Chem.* 4, 62, 1954.
110. Pantzer, R. *Farbe Lack* 84, 999, 1978.
111. Svoboda, M., and J. Mleziva. *Prog. Org. Coat.* 12, 251, 1984.
112. Rosenfeld, I.L. *Zashch. Met.* 15, 349, 1979.
113. Largin, B.M., and I.L. Rosenfeld. *Zashch. Met.* 17, 408, 1981.

5 Waterborne Coatings

Most of the important types of modern solvent-borne coatings—acrylics, epoxies, alkyds, polyurethanes, and polyesters—are also available in waterborne formulations. Waterborne coatings are mainly used for decorative purposes, but two-component, heavy-duty protective waterborne coatings are also available [1]. Acrylics are dominating the waterborne market and are several times larger in volume than all the other technologies together.

Waterborne paints are not simply solvent-borne paints in which the organic solvent has been replaced with water; the paint chemist must design an entirely new system from the ground up. In this chapter, we discuss how waterborne paints differ from their solvent-borne counterparts.

Waterborne paints are by nature more complex and more difficult to formulate than solvent-borne coatings. The extremely small group of polymers that are soluble in water does not, with a few exceptions, include any that can be usefully used in paint. In broad terms, a one-component, solvent-borne coating consists of a polymer dissolved in a suitable solvent. Film formation consists of merely applying the film and waiting for the solvent to evaporate. In a waterborne coating, the polymer particles are not at all dissolved; instead, they exist as solid polymer particles dispersed in the water. Film formation is more complex when wetting, thermodynamics, and surface energy theory come into play. Among other challenges, the waterborne paint chemist must

- Design a polymer reaction to take place in water so that monomer building blocks polymerize into solid polymer particles
- Find additives that can keep the solid polymer particles in a stable, even dispersion, rather than in clumps at the bottom of the paint can
- Find more additives that can somewhat soften the outer part of the solid particles, so that they flatten more easily during film formation

And all this was just for the binder. Additional specialized additives are needed, for example, to keep the pigment from clumping; these are usually different for dispersion in a polar liquid, such as water, than in a nonpolar organic solvent. The same can be said for the chemicals added to make the pigments integrate well with the binder, so that gaps do not occur between the binder and pigment particles. And, of course, more additives unique to waterborne formulations may be used to prevent flash rusting of the steel before the water has evaporated. (It should perhaps be noted that the need for flash rusting additives is somewhat questionable.)

5.1 TECHNOLOGIES FOR POLYMERS IN WATER

Most polymer chains are not polar; water, being highly polar, cannot dissolve them. Chemistry, however, has provided ways to get around this problem. Paint technology has taken several approaches to suspending or dissolving polymers in water. All of them require some modification of the polymer to make it stable in a water dispersion or solution. The concentration of the polar functional groups plays a role in deciding the form of the waterborne paint: a high concentration confers water solubility, whereas a low concentration leads to dispersion [2]. Much research has been ongoing to see where and how polar groups can be introduced to disrupt the parent polymer as little as possible.

5.1.1 WATER-REDUCIBLE COATINGS AND WATER-SOLUBLE POLYMERS

In both water-reducible coatings and water-soluble polymers, the polymer chain, which is naturally hydrophobic, is altered; hydrophilic segments such as carboxylic acid groups, sulfonic acid groups, and tertiary amines are grafted onto the chain to confer a degree of water solubility.

In water-reducible coatings, the polymer starts out as a solution in an organic solvent that is miscible with water. Water is then added. The hydrophobic polymer separates into colloid particles, and the hydrophilic segments stabilize the colloids [3]. Water-reducible coatings, by their nature, always contain a certain fraction of organic solvent.

Water-soluble polymers do not begin in organic solvent. These polymers are designed to be dissolved directly in water. An advantage to this approach is that drying becomes a much simpler process because the coating is neither dispersion nor emulsion. In addition, temperature is not as important for the formation of a film with good integrity. The polymers that lend themselves to this technique, however, are of lower molecular weight (10^3 to 10^4) than the polymers used in dispersions (10^5 to 10^6) [4].

5.1.2 AQUEOUS EMULSION COATINGS

An emulsion is a dispersion of one liquid in another; the best-known example is milk, in which fat droplets are emulsified in water. In an emulsion coating, a liquid polymer is dispersed in water. Many alkyd and epoxy paints are examples of this type of coating.

5.1.3 AQUEOUS DISPERSION COATINGS

In an aqueous dispersion coating, the polymer is not water soluble at all. Rather, it exists as a dispersion or latex of very fine (50–500 nm diameter) solid particles in water. It should be noted that merely creating solid polymer particles in organic solvent, removing the solvent, and then adding the particles to water does not produce aqueous dispersion coatings. For these coatings, the polymers must be produced in water from the start. Most forms of latex begin as emulsions of the polymer building blocks and then undergo polymerization. Polyurethane dispersions, on the other hand, are produced by polycondensation of aqueous building blocks [3].

5.2 WATER VERSUS ORGANIC SOLVENTS

The difference between solvent-borne and waterborne paints is due to the unique character of water. In most properties that matter, water differs significantly from organic solvents. In creating a waterborne paint, the paint chemist must start from scratch, reinventing almost everything from the resin to the last stabilizer added.

Water differs from organic solvents in many aspects. For example, its dielectric constant is more than an order of magnitude greater than those of most organic solvents. Its density, surface tension, and thermal conductivity are greater than those of most of the commonly used solvents. For its use in paint, however, the following differences between water and organic solvents are most important:

- *Water does not dissolve the polymers that are used as resins in paints.* Consequently, the polymers have to be chemically altered so that they can be used as the backbones of paints. Functional groups, such as amines, sulfonic groups, and carboxylic groups, are added to the resins to make them soluble or dispersible in water.
- *Water has a high freezing point compared with organic solvents.* Waterborne paint must be stored and applied under frost-free conditions.
- *The latent heat of evaporation is lower for water than for many organic solvents used in paint.* Waterborne paints therefore typically have shorter drying times than solvent-borne ones.
- *The surface tension of water is higher than that of the solvents commonly used in paints.* This high surface tension plays an important part in the film formation of latexes (see Section 5.3).

Waterborne coatings always contain some organic solvent (cosolvent), typically in the order of 5%–10%. The cosolvent tends to be low-molecular-weight ketones, alcohols, and esters. Their function in the paint is to help the film formation process.

5.3 LATEX FILM FORMATION

Waterborne dispersions form films through a fascinating process. In order for cross-linking to occur and a coherent film to be built, the solid particles in dispersion must spread out as the water evaporates. They will do so because coalescence is thermodynamically favored over individual polymer spheres: the minimization of the total surface allows for a decrease in free energy [5].

Film formation can be described as a three-stage process. The stages are described below; stages 1 and 2 are depicted in Figure 5.1.

1. *Colloid concentration.* The bulk of the water in the newly applied paint evaporates. As the distance between the spherical polymer particles shrinks, the particles move and slide past each other until they are densely packed.

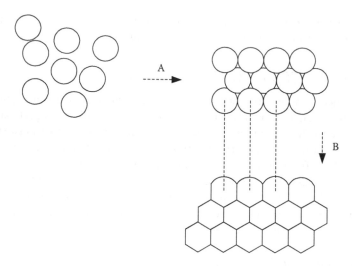

FIGURE 5.1   Latex film formation: colloid concentration (a) and coalescence (b). Note that center-to-center distances between particles do not change during coalescence.

The particles are drawn closer together by the evaporation of the water but are themselves unaffected; their shape does not change.

2. *Coalescence*. This stage begins when the only water remaining is in between the particles. In this second stage, also called the "capillary stage," the high surface tension of the interstitial water becomes a factor. The water tries to reduce its surface at both the water–air and water–particle interfaces. The water actually pulls enough on the solid polymer particles to deform them. This happens on the sides, above, and below the sphere; everywhere it contacts another sphere, the evaporating water pulls it toward the other sphere. As this happens on all sides and to all spheres, the result is a dodecahedral honeycomb structure. Organic cosolvents help the coalescence.

3. *Macromolecule interdiffusion*. Under certain conditions, such as sufficiently high temperatures, the polymer chains can diffuse across the particle boundaries. A more homogeneous, continuous film is formed. Mechanical strength and water resistance of the film increase [5,6].

5.3.1   DRIVING FORCE OF FILM FORMATION

The film formation process is extremely complex, and there are a number of theories—or more accurately, schools of theories—to describe it. A major point of difference among them is the driving force for particle deformation: surface tension of the polymer particles, Van der Waals attraction, polymer–water interfacial tension, capillary pressure at the air–water interface, or combinations of the above. These models of the mechanism of latex film formation are necessary in order to improve existing waterborne paints and design the next generation. To improve the rate of film formation, for example, it is important to know if the main driving force for coalescence is located at the interface between the polymer and water, between

TABLE 5.1
Estimates of Forces Operating during Particle Deformation

Type of Force Operating	Estimated Magnitude (N)
Gravitational force on a particle	6.4×10^{-17}
Van der Waals force (separation 5 nm)	8.4×10^{-12}
Van der Waals force (separation 0.2 nm)	5.5×10^{-9}
Electrostatic repulsion	2.8×10^{-10}
Capillary force due to receding water–air interface	2.6×10^{-7}
Capillary force due to liquid bridges	1.1×10^{-7}

Source: Visschers, M., et al., *Prog. Org. Coat.*, 31, 311, 1997. With permission from Elsevier.

water and air, or between polymer particles. This location determines which surface tension or surface energies should be optimized.

In recent years, a consensus seems to be growing that the surface tension of water, at either the air–water or polymer–water interface, or both—is the driving force. Atomic force microscopy (AFM) studies seem to indicate that capillary pressure at the air–water interface is most important [7]. Working from another approach, Visschers and colleagues [8] have reported supporting results. They estimated the various forces that operate during polymer deformation for one system, in which a force of 10^{-7} N would be required for particle deformation. The forces generated by capillary water between the particles and by the air–water interface are both large enough (Table 5.1).

Gauthier and colleagues have pointed out that polymer–water interfacial tension and capillary pressure at the air–water interface are expressions of the same physical phenomenon and can be described by the Young and Laplace laws for surface energy [5]. The fact that there are two minimum film formation temperatures (MFFTs), one "wet" and one "dry," may be an indication that the receding polymer–water interface and evaporating interstitial water are both driving the film formation (see Section 5.4).

For more in-depth information on the film formation process and important thermodynamic and surface energy considerations, consult the excellent reviews by Lin and Meier [7], Gauthier et al. [5], or Visschers et al. [9]. All these reviews deal with nonpigmented latex systems. The reader working in this field should also become familiar with the pioneering works of Brown [10], Mason [11], and Lamprecht [12].

5.3.2 HUMIDITY AND LATEX CURE

Unlike organic solvents, water exists in the atmosphere in vast amounts. Researchers estimate that the atmosphere contains about 6×10^{15} L of water [13,14]. Because of this fact, relative humidity is commonly believed to affect the rate of evaporation of water in waterborne paints. Trade literature commonly implies that waterborne coatings are somehow sensitive to high-humidity conditions. However, Visschers et al. have elegantly shown this belief to be wrong. They used a combination of thermodynamics and contact-angle theory to prove that latex paints dry at practically all humidities as long as they are not directly wetted—that is, by rain or condensation [8]. Their results have been borne out in experiments by Forsgren and Palmgren [15], who found that changes in relative humidity had no significant effect on the mechanical and physical properties of the cured coating. Gauthier and colleagues have also shown experimentally that latex coalescence does not depend on ambient humidity. In studies of water evaporation using weight loss measurements, they found that the rate in stage 1 depends on ambient humidity for a given temperature. In stage 2, however, when coalescence occurs, the water evaporation rate could not be explained by the same model [5].

5.3.3 REAL COATINGS

The models for film formation described above are based on latex-only systems. Real waterborne latex coatings contain much more: pigments of different kinds; coalescing agents to soften the outer part of the polymer particles; and surfactants, emulsifiers, and thickeners to control wetting and viscosity and maintain dispersion.

Whether a waterborne paint will succeed in forming a continuous film depends on a number of factors, including

- Wetting of the polymer particles by water (Visschers and colleagues found that the contact angle of water on the polymer sphere has a major influence on the contact force that pushes the polymer particles apart [if positive] or pulls them together [if negative] [8])
- Polymer hardness
- Effectiveness of the coalescing agents
- Ratio of binder to pigment
- Dispersion of the polymer particles on the pigment particles
- Relative sizes of pigment to binder particles in the latex

5.3.3.1 Pigments

To work in a coating formulation, whether solvent-borne or waterborne, a pigment must be well dispersed, coated by a binder during cure, and in the proper ratio to the binder. The last point is the same for solvent-borne and waterborne formulations; however, the first two require consideration in waterborne coatings.

The high surface tension of water affects not only polymer dispersion but also pigment dispersion. As Kobayashi has pointed out, the most important factor in dispersing a pigment is the solvent's ability to wet it. Because of surface tension considerations, wetting depends on two factors: hydrophobicity (or hydrophilicity) of the pigment and the pigment geometry. The interested reader is directed to Kobayashi's review for more information on pigment dispersion in waterborne formulations [16].

Joanicot and colleagues examined what happens to the film formation process described above when pigments much larger in size than latex particles are added to the formulation. They found that waterborne formulations behave similarly to solvent-borne formulations in this matter: the pigment volume concentration (PVC) is critical. In coatings with low PVC, the film formation process is not affected by the presence of pigments. With high PVC, the latex particles are still deformed as water evaporates but do not exist in sufficient quantity to spread completely over the pigment particles. The dried coating resembles a matrix of pigment particles that are held together at many points by latex particles [17].

The problems of PVC pigment dispersion imbalance are shown in Figure 5.2.

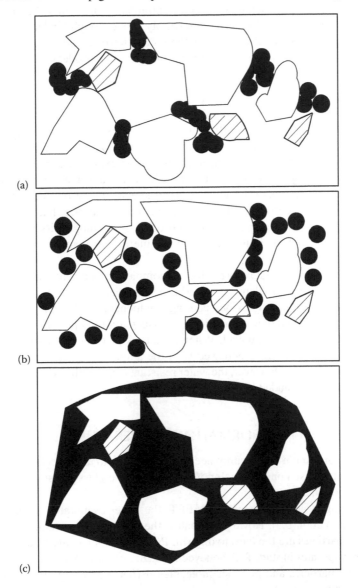

(a)

(b)

(c)

FIGURE 5.2 Pigment and binder particle combinations. The polymer particles are black, and the pigment particles are white or striped (representing two different pigments). (a) High PVC, with binder particles aggregated between pigment particles. (b) High PVC and dispersed binder particles. (c) Low PVC and enough binder to fill all gaps between pigment particles.

In Figure 5.2a, the PVC is very high and the binder particles have flocculated at a limited number of sites between pigment particles. When they deform, the film will consist of pigment particles held together in places by polymer, with voids throughout.

Figure 5.2b shows the same very high PVC, but here the binder particles are dispersed. The binder particles may form a continuous film around the pigment particles, but voids still occur because there simply is not enough binder.

Figure 5.2c shows the ideal scenario: the PVC is lower, and the surrounding black binder is able to not only cover the pigment particles but also leave no void between them.

5.3.3.2 Additives

In real waterborne paints, the film formation process can result in a nonhomogeneous layer of cured paint. Tzitzinou and colleagues, for example, have shown that the composition of a cured paint layer can be expected to vary through the depth of the coating. They studied an anionic surfactant in an acrylic latex film. Using AFM and Rutherford backscattering spectrometry on cured films, they found a higher concentration of surfactant at the air surface than in the bulk of the coating [18]. Wegmann has also studied the inhomogeneity of waterborne films after cure, but attributes his findings mainly to insufficient coalescence during cure [19].

The chemistry of real latex formations is complex and currently defies predictive modeling. A reported problem for waterborne modelers is that an increase in curing temperature can affect various coating components differently. Snuparek and colleagues added a nonionic emulsifier to a dispersion of copolymer butyl methacrylate–butyl acrylate–acrylic acid. When cure took place at room temperature, the water resistance of the films increased with the amount of emulsifier added. When cure happened at 60°C, however, the water resistance of the films *decreased* with the amount of emulsifier added [4].

5.4   MINIMUM FILM FORMATION TEMPERATURE

MFFT is the minimum temperature needed for a binder to form a coherent film. This measurement is based on, although not identical to, the glass transition temperature (T_g) of the polymer.

If a coating is applied below the MFFT, the water evaporates as described for stage 1 (see Section 5.3). However, because the ambient temperature is below the MFFT, the particles are too hard to deform. Particles do not coalesce as the interstitial water evaporates in stage 2. A honeycomb structure, with Van der Waals bonding between the particles and polymer molecules diffused across particle boundaries, does not occur.

The MFFT can be measured in the laboratory as the minimum temperature at which a cast latex film becomes clear. This is simply because if the coating has not formed a coherent film, it will contain many voids between polymer particles. These voids create internal surfaces within the film, which cause the opacity.

Latexes must always be applied at a temperature above the MFFT. This is more difficult than it sounds, because the MFFT is a dynamic value, changing over time. In a two-component system, the MFFT begins increasing as soon as the components are mixed. Two-component waterborne paints must be applied and dried before the MFFT has increased enough to reach room temperature. When the MFFT has reached room temperature and the end of pot life has been reached for a waterborne paint, viscosity does not increase as it does with many solvent-borne paints [20]. Hence, the applicator cannot see that pot life is exceeded from the viscosity of the paint, but must watch the time.

5.4.1 WET MFFT AND DRY MFFT

If a latex paint is dried below the MFFT, then no particle deformation occurs. If the temperature of the dried (but not coalesced) latex is then raised to slightly above the MFFT, no coalescence as described in Section 5.3 should occur either; no receding air–water interface exists to generate capillary forces, and thus no particle deformation occurs. If the temperature is further raised, however, particle deformation eventually occurs. This is because some residual water is always left between the particles due to capillary condensation. At the higher temperature, these liquid bridges between the particles can exert enough force to deform the particles.

Two MFFTs appear to exist: wet MFFT and dry MFFT. The normal, or wet, MFFT is that which is seen under normal circumstances—wherein a latex is applied at an ambient temperature above the polymer's T_g, and film formation follows the three stages described in Section 5.3. This wet MFFT is associated with particles deforming due to a receding air–water interface.

The higher temperature at which a previously uncoalesced latex deforms is the dry MFFT. This is associated with much smaller quantities of water between particles. The role of the water at this higher dry MFFT is not well understood. It may be that the smaller amounts are able to deform the particles because a different deformation mechanism is possible at the elevated temperature. Or, it may be that the polymer particle is softer under these circumstances. The phenomenon is interesting and may be helpful in improving models of latex film formation [21–24].

5.5 FLASH RUSTING

Nicholson defines flash rusting as "the rapid corrosion of the substrate during drying of an aqueous coating, with the corrosion products (i.e., rust) appearing on the surface of the dried film" [25]. Flash rusting is commonly named as a possible drawback to waterborne coatings; yet, as Nicholson goes on to point out, the phenomenon is not understood and its long-term importance for the coating is unknown. Studies have been carried out to identify effective anti-flash-rust additives; however, because they are empirical in approach, the mechanisms by which any of them work—or even the necessity for them—has not been well defined.

The entire flash rusting discussion may be unnecessary. Igetoft [26] has pointed out that flash rusting requires not only water but also salt to be present. The fact that steel is wet does not necessarily mean that it will rust.

Forsgren and Persson [27] obtained results that seem to indicate that flash rusting is not a serious problem with modern waterborne coatings. They used contact-angle measurements, Fourier transform infrared (FTIR) spectroscopy, and AFM to study changes in surface chemistry at the steel–waterborne acrylic coating interface before curing takes place. In particular, the total free surface energy of the steel and its electromagnetic and acid–base components were studied before and immediately after application of the coating. The expectation was that the acidic or basic components, or both, of the steel's surface energy would increase immediately after the coatings were applied. Instead, the total surface energy of the steel decreased, and the Lewis base component dropped dramatically. The contact-angle measurements after contact with the coatings were more typical of polymers than of cold-rolled steel. Spectroscopy studies showed carboxyl and alkane groups on the surface of the steel after two minutes' exposure to the paint. AFM showed rounded particles of a softer material than steel distributed over the surface after a short exposure to the coatings. The authors speculated that the adhesion promoters on the polymer chain are so effective that the first particles of polymer are already attached to the steel after 20 seconds—in other words, before any deformation due to water evaporation could have occurred. The effects of this immediate bonding on immediate and long-term corrosion protection are unknown. Better knowledge of the processes taking place at the coating–metal interface immediately upon application of the coating may aid in understanding and preventing undesirable phenomena such as flash rusting.

REFERENCES

1. Hawkins, C.A., A.C. Sheppard, and T.G. Wood. *Prog. Org. Coat.* 32, 253, 1997.
2. Padget, J.C. *J. Coat. Technol.* 66, 89, 1994.
3. Misev, T.A. *J. Jpn. Soc. Colour Mater.* 65, 195, 1993.
4. Snuparek, J., et al. *J. Appl. Polym. Sci.* 28, 1421, 1983.
5. Gauthier, C., et al. *Film Formation in Water-Borne Coatings*, ed. T. Provder, M.A. Winnik, and M.W. Urban. ACS Symposium Series 648. Washington, DC: American Chemical Society, 1996.
6. Gilicinski, A.G., and C.R. Hegedus. *Prog. Org. Coat.* 32, 81, 1997.
7. Lin, F., and D.J. Meier. *Prog. Org. Coat.* 29, 139, 1996.
8. Visschers, M., J. Laven, and R. van der Linde. *Prog. Org. Coat.* 31, 311, 1997.
9. Visschers, M., J. Laven, and A.L. German. *Prog. Org. Coat.* 30, 39, 1997.
10. Brown, G.L. *J. Polym. Sci.*, 22, 423, 1956.
11. Mason, G. *Br. Polym. J.* 5, 101, 1973.
12. Lamprecht, J. *Colloid Polym. Sci.* 258, 960, 1980.
13. Nicholson, J. *Waterborne Coatings*. Oil and Colour Chemists' Association Monograph 2. London: Oil and Colour Chemists' Association, 1985.
14. Franks, F. *Water*. London: Royal Society of Chemistry, 1983.
15. Forsgren, A., and S. Palmgren. Effect of application climate on physical properties of three waterborne paints. Report 1997:3E. Stockholm: Swedish Corrosion Institute, 1997.
16. Kobayashi, T. *Prog. Org. Coat.*, 28, 79, 1996.
17. Joanicot, M., V. Granier, and K. Wong. *Prog. Org. Coat.*, 32, 109, 1997.
18. Tzitzinou, A., et al. *Prog. Org. Coat.* 35, 89, 1999.
19. Wegmann, A. *Prog. Org. Coat.* 32, 231, 1997.

20. Nysteen, S. Personal communication. Hempel's Marine Paints A/S (Denmark).
21. Sperry, P.R., et al. *Langmuir* 10, 2169, 1994.
22. Keddie, J.L., et al. *Macromolecules* 28, 2673, 1995.
23. Snyder, B.S., et al. *Polym. Preprints* 35, 299, 1994.
24. Heymans, D.M.C., and M.F. Daniel. *Polym. Adv. Technol.* 6, 291, 1995.
25. Nicholson, J.W. The widening world of surface coatings, in *Surface Coatings*, ed. A.D. Wilson, J.W. Nicholson, and H.J. Prosser. Amsterdam: Elsevier Applied Science, 1988, chap. 1.
26. Igetoft, L. Våtblästring som förbehandling före rostskyddsmålning—litterature-genomgång. Report 61132:1. Stockholm: Swedish Corrosion Institute, 1983.
27. Forsgren, A., and D. Persson. Changes in the surface energy of steel caused by acrylic waterborne paints prior to cure. Report 2000:5E. Stockholm: Swedish Corrosion Institute, 2000.

6 Powder Coatings

Powder coatings are the youngest of the coatings technologies, but also the fastest growing. The first powder coatings were developed in the 1950s, based on polyethylene powders applied by a fluidized bed technology on preheated steel. In the 1960s, powder coatings based on polyesters and epoxies, as we know them today, were developed, and the use of powder coatings started to grow. Powder coatings constitute about 11% of the world coating market today [1].

In a powder coating, all the constituents of the coating, binder, fillers, pigments, and additives are present in a dry, solventless powder. The powder is applied to the metal substrate by spraying or in a fluidized bed. When spraying the powder, the particles move through a gun where they receive an electrostatic charge, and this charge draws them to the metal substrate that is electrically connected to ground. The layer of powder is then melted in an oven at a temperature between 160°C and 210°C, where it floats together into a continuous film. In the fluidized bed, the object to be coated is preheated above the melting temperature of the powder and dipped in the fluidized bed, and the powder simply melts and sticks to the object.

Due to environmental concern and legislation, the solvent-free nature of powder coatings has been a strong argument in their favor. The increasing focus on solvent emission is an important reason why powder coatings have been and still are the fastest-growing coating technology. From a solvent emission perspective, powder coatings are the ultimate technology, having virtually zero emission of volatile organic compounds (VOCs). However, their success is also related to a very beneficial performance-to-cost ratio. Powder coatings have a number of technical and economic benefits, for example,

- Since the powder is drawn to the substrate by their electrostatic charge, a very high ratio of the powder actually ends up on the substrate. When spraying conventional liquid coatings, a considerable fraction of the paint does not hit the substrate and is lost, depending on substrate geometry. In addition, the powder that does not hit the substrate is recycled in the application booth and sprayed again. This both decreases waste and improves the cost-effectiveness of powder coatings, compared with conventional liquid coatings.
- The high-temperature curing of the powder takes only a few minutes to complete, and after cooling the coated item can be used immediately. Compared with conventional wet paint, where solvent evaporation and curing at ambient temperature will take days, powder coatings are ready for use when the product leaves the coating line.

- Since powder coatings are applied more or less automatically in the production line, their cost-effectiveness often compares favorably to other coatings.
- Curing at high temperature results in a very dense and highly protective film.

The most important limitation to the use of powder coatings is the size of the objects to be coated. They must fit in the powder application line. Although fairly large objects can be powder coated, the technology is limited to objects and parts. Powder coating of nonconducting materials, such as wood, polymers, and ceramics, is also less trivial, but solutions have been developed for powder coating of such substrates as well. For example, medium-density fiberboard (MDF) boards used in walls and furniture are often powder coated. However, more than 90% of powder coatings are applied on metals. Powder coating of nonmetallic materials will not be covered in this book, since our subject is anticorrosion coatings.

6.1   GENERIC TYPES OF POWDER COATINGS AND RANGE OF USE

The most common polymers used in powder coatings are polyester, epoxy, polyurethane, polyester–epoxy mix (hybrid), and acrylics. Like conventional paint, powder coatings can be thermoplastic or thermosetting. During the film formation process of thermoplastic powder coatings, the powder particles melt and flow together into a continuous film of entangling polymer chains. The film may be remelted at high temperature or dissolved by a solvent, parallel to a physically drying wet paint. In thermosetting powder coatings, the two reactants are both mixed into the powder during production of the powder, but their reaction rate is so low at ambient temperature that the powder is stable during storage. The curing reaction first starts at a temperature well above the melting point of the powder, after the powder has formed a film.

6.1.1   THERMOPLASTIC POWDER COATINGS

The first powder coatings were based on thermoplastic polymers; nowadays, however, thermoplastic coatings only constitute a small fraction of the powder coating market. The thermoplastic coatings generally have weaker adhesion, which means that they often need a primer coat. They also generally have less good barrier properties than the thermosets, and must be applied in high thickness, often several hundred micrometers, in order to be protective. Most of them have a higher melting temperature than thermosetting powders, and therefore require baking at higher temperature to melt and form a film. However, the various thermoplastics also offer some distinguished properties that make them useful in many applications. The most important thermoplastic powder coatings are vinyls, nylons, and polyolefins. Historically, thermoplastic powder coatings have been applied by fluidized bed, and that is still the case for many of these coatings.

Vinyls include polyvinyl chloride (PVC) and polyvinylidene fluoride (PVDF), which are the two most important thermoplastic vinyls. PVC is a brittle and hard polymer, and in order to make a sufficiently flexible film, plasticizers must be added.

PVC is degraded by ultraviolet (UV) light and has low durability outdoors. PVC coatings are therefore mainly applied on products for indoor use. They are mechanically tough, protective coatings when applied in sufficient thickness and are therefore, for example, used for the protection of dishwasher baskets. They are also approved for food contact. PVC is an inexpensive material, which makes the coatings affordable. The low price per kilogram powder compensates for the high thickness needed, and the price per square meter may end up at the same level as that of other coatings. In order to get the high film thickness needed, PVC is often applied by fluidized bed technology. In spite of belonging to the same group of polymers, PVDF-based powder coatings have very different properties from PVC and different applications. They have excellent UV resistance and gloss retention, and good abrasion resistance. Being a fluoropolymer, they have low surface energy, which makes them somewhat dirt repellent. These properties make them excellent topcoats and well suited as decorative coatings for architectural applications.

Nylon-based powder coatings are mechanically tough, tolerate high temperature, and resist a wide range of chemicals and solvents. They are therefore frequently used for the coating of handles, wire goods, automotive parts, and medical equipment. They are an alternative to PVC coatings and are used for many of the same products, but offer even stronger resistance against mechanical wear and have somewhat better UV resistance. Nylon coatings are also approved for contact with food.

Polyolefins, that is, polyethylene and polypropylene, produce coatings with a smooth surface and a soft, almost waxy feel. As inert polymers without functional groups, they have almost no water absorption and are very resistant to chemicals and solvents. Due to these benefits, polyolefin coatings are frequently used on lab equipment that require regular washing. They are usually chosen for their low cost and ease of application. Their inert nature and lack of functional groups also mean that they have very poor adhesion to metals, and primers or adhesion promoters are often used. Like many of the other thermoplastic powder coatings, they are often applied in a fluidized bed.

6.1.2 THERMOSETTING POWDER COATINGS

As stated above, the thermosetting powders dominate the market, and polyesters, epoxies, and polyester–epoxy hybrids are the dominating thermosetting powders. The epoxies are used for heavy-duty corrosion protection, while the polyesters are used for both decorative and protective purposes outdoors. The polyester–epoxy hybrids combine the decorative properties of the polyester with the toughness of the epoxy, and are the preferred binder for indoor exposure. The thermosetting binders crosslink in a curing reaction, which turns the film into a three-dimensional network of virtually a single enormous molecule. The properties of the binder depend on the nature of the polymer and crosslink density. The crosslink density in the cured coating depends on the functionality of the resin and the curing agent, that is, the number of reactive sites.

Epoxy powder coatings have strong adhesion, are very resistant to chemicals, have strong wear resistance, and provide excellent corrosion resistance. However, like wet epoxy paint, they are not resistant to UV light and cannot be exposed outdoors without chalking or yellowing, which is the most important limitation to their use. The most frequently used epoxy monomer in powder coatings is bisphenol A, but bisphenol F and epoxy novalacs are also used, just as in the wet epoxy paints. Typical curing agents are phenols, dicyandiamide (DICY), aromatic amines, and aliphatic diamines.

Epoxies are typically used for corrosion protection and less for decorative purposes. Pipelines, rebar, and rock bolts are examples of steel products that are coated with epoxy powder coatings. The largest market for epoxy powder coatings is onshore and offshore oil and gas pipelines. The application of the powder to the pipe spools differs from other powder application lines, due to the large size of the spools, which will be explained in Section 6.3. In the pipeline coating industry, epoxy powder coatings go under the name fusion-bonded epoxy (FBE). The term *fusion bonded* refers to the crosslinking of the binder. The FBE coating is often the first coat in a multicoat system, but it can also be used alone. Kehr has written a comprehensive review on the use of FBE for corrosion protection of pipelines [2].

Polyester powder coatings also have good adhesion and protective properties, but contrary to the epoxies, they are quite resistant to UV light and can be exposed to sunlight. Polyesters are therefore used for almost everything that will be exposed outdoors. Primarily two curing agents have been used for polyester powders, triglycidyl isocyanurate (TGIC) and primid. TGIC has historically been the dominating curing agent, but is now being replaced by primid due to its classification as toxic, irritating, sensitizing, mutagenic, and dangerous for the environment. The primid is safer to use, and now completely dominates the market in Europe. In the rest of the world, the transition from TGIC to primid has just started. A minor drawback with primid is that water is produced in the curing reaction. This water must leave the film during the curing process to avoid formation of pores in the film. This limits the maximum film thickness of primid-cured polyesters. The chance for formation of pinholes may also increase slightly. Typical products coated with polyesters include architectural components such as facade panels, door and window frames, automobile and bicycle parts, agricultural equipment, household appliances, and electrical enclosures.

As mentioned above, the hybrid powder coatings based on a mixture of polyester and epoxy resins are the preferred coatings for indoor exposure, commonly applied to furniture and kitchen appliances. The epoxy is primarily added to improve the mechanical properties of the polyester. Epoxy–polyester hybrid powders are not more weather resistant than epoxies. They are still susceptible to chalking and will very quickly lose their gloss if exposed outdoors.

Acrylic powder coatings are used primarily in the automotive sector as clear topcoat, giving a very smooth and shiny finish. They produce a very hard and chip-resistant coating. The binder is based on glycidyl methacrylate (GMA). Acrylics can also be added to polyesters to make polyester–acrylic hybrid coatings, where the acrylic portion will promote improved flow and leveling, as well as enhanced stain and chemical resistance. Acrylics are also used as crosslinkers for polyesters, where they will have an additional effect as gloss-control agents, making decorative and matte coatings.

6.2   POWDER PRODUCTION

The production of the powder for powder coatings has more in common with the production of plastics than with the production of solvent or waterborne coatings. A production line is outlined in Figure 6.1. The process can be either discontinuous, that is, in batches, or continuous, but the process steps will be the same:

- Weighing (1)
- Premixing (2)
- Melting, kneading, and extrusion (3)
- Cooling and crushing (4)
- Grinding (5)
- Particle size classification (6)
- Packaging (7)

A thorough premixing of the powder is important for good distribution of the components and homogeneity of the powder. After premixing, the powder enters the extruder, where it is melted and kneaded with further homogenization of the components. The melt temperature is typically around 120°C, where the reaction between the resin and curing agent in a thermosetting powder is still rather slow. The mixture is then extruded and cooled to a solid material. This is first crushed to pieces and then grinded into a fine powder. The grinder produces a distribution of particle sizes that is somewhat wider than the particle size distribution wanted for the powder. A separation of the powder, by either sieving or centrifugation, taking out the too-small and too-large particles is required. The classified powder goes to packaging, while the rest is recycled back into the production.

Particle size is a critical parameter in powder coatings, determining both powder application and film formation properties [3]. A certain size distribution is desirable, but particles below 10 μm are generally not wanted. The maximum particle size typically varies between 40 and 100 μm, while the median particle size, denoted

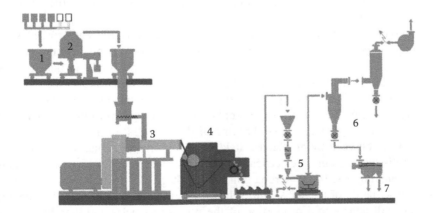

FIGURE 6.1   Production of the powder.

X_{50}, is normally in the range between 30 and 50 μm. The effect of particle size on application and film formation is discussed in the following sections.

6.3 APPLICATION TECHNOLOGY

Application of the powder is of course very different from applying a wet paint. Four techniques are used: electrostatic spraying, fluidized bed, flocking gun, and flame spraying, where electrostatic spraying is by far the most common.

6.3.1 ELECTROSTATIC SPRAYING

The basic principle of electrostatic spraying is that the powder particles are blown through the application gun by means of compressed air, where it receives an electric charge. The substrate is electrically connected to ground through the hangers, so that the charged particles are drawn to the substrate by an electrostatic force. The airflow out of the gun also gives them a velocity toward the substrate. Consisting mainly of a polymer, the powder particles are electrically insulating and retain much of their electric charge when they hit the substrate, which makes them stick to the substrate. The film thickness is somewhat self-limiting. The accumulation of charged particles on the surface will have a shielding effect, and the deposition of new particles will slow down, directing them to areas with less powder. This effect contributes to producing a uniform film thickness.

Particles that do not hit or stick to the substrate will fall down to the bottom of the application booth and be recycled. The volume fraction of sprayed powder that ends up on the coated object is called transfer efficiency. The transfer efficiency will depend on the geometry of the object. Large objects such as facade panels typically can result in about 80% transfer efficiency, while small surfaces and objects with lots of open space, such as a bicycle frame, can attain only about 25%. Particles that receive little charge will to a smaller extent be deposited on the substrate and end up in the powder recycling. Hence, the recycled powder will have an increased concentration of particles with poor charging properties. There is therefore a limit to powder recycling. At some point, the powder in the recycling should be discarded.

Two types of powder application guns are used, with different principles for charging the powder: tribo and corona (Figure 6.2). In the tribo gun, the powder particles are charged by the friction between the particles and the powder tube inside the gun. The particles receive a positive charge, which draws the particles to the grounded object. Only powders developed for the tribo gun can be applied. However, most of them can also be applied by corona guns. Compared with the corona gun, tribo guns give less charging of the particles. In the corona gun, the particles are charged by an electric field. In electricity, a corona discharge is an electric discharge from an electrically charged surface brought about by ionizing the surrounding gas. In the corona gun, high voltage (30–100 kV) ionizes the air inside or at the tip of the spray gun. When the powder passes through this ionized air, ions will adhere to a proportion of the powder particles and thereby apply a

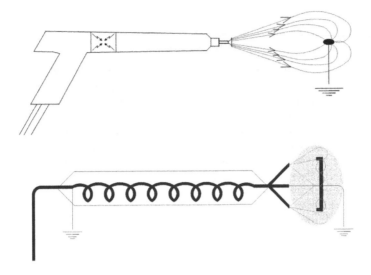

FIGURE 6.2   Powder application guns. Upper: Corona. Lower: Tribo.

negative charge to them. Even though the voltage in the gun is high, the current is very low, so the energy input is quite low.

Since the powder is applied with an electrostatic charge, there will be an electric field between the spray gun and the substrate. On a flat substrate, a rather homogeneous electric field is formed, resulting in fairly homogeneous distribution of the powder. However, on a shaped work piece, field lines will tend to concentrate on the points closest to the spray gun, so that these receive more powder than surfaces farther away. This is called the Faraday cage effect. This effect typically causes problems with low film thickness on internal corners and remote areas. A beneficial effect of the electric field is the "wraparound effect." On small objects, both the front and the backside can be coated by spraying from one side. The electric field will draw the powder to the backside of the object. Table 6.1 presents advantages and disadvantages with both methods [4].

The particle size distribution of the powder will affect several aspects of the coating application process:

- Transport of the powder in the application equipment: Particles below 10 μm cause flow problems. The small particles fill the voids between the larger particles, and the powder becomes packed and difficult to fluidize. A higher air pressure is needed to press the powder through the system. The powder flow becomes less uniform, resulting in film thickness variation and other problems.
- Electrostatic charging of the particles: Smaller particles receive more charge per weight than the larger particles, which affects their flow pattern to the substrate. The distribution of particle sizes on the substrate may not be homogeneous.

TABLE 6.1
Comparison of Corona and Tribo Methods

	Advantages	Disadvantages
Corona	• Rapid and strong charging of the particles • Strong electrostatic field lines move the particles effectively to the object • Repairs of the powdered surface are possible • Light and robust spray gun • Accepts different types of powder materials and particle sizes • Film thickness can be changed by adjusting the voltage	• Self-limiting effect due to strong charging of the particles limits maximum film thickness • Strong field lines lead to Faraday cage effect; i.e., internal corners and crevices may not be properly covered
Tribo	• Little Faraday effect; crevices, internal corners, and hollow spaces are better penetrated • Powder can be better directed by the use of directional finger sprayers and aerodynamics • Uniform coating • No high-voltage generator required • Higher film thickness possible due to smaller self-limiting effect	• Uncontrolled airstreams have more effect on application • Special powder is necessary; formulation must be adapted to the tribo charging process • Particles smaller than 10 microns are difficult to charge • Powder deposition slower than for corona and more guns are required • Higher gun investment cost for equal capacity output • More wear and shorter lifetime of gun internals and other parts

Source: Thies, M.J., Comparison of tribo and corona charging methods: How they work and the advantages of each, in *Powder Coating '94*, Powder Coating Institute, Taylor Mill, KY, 1994, pp. 235–251. With permission from Powder Coating Institute.

- Faraday cage penetration: Smaller particles will have a higher charge-to-weight ratio and will give a stronger Faraday cage effect.
- Transfer efficiency: Particles below 25 μm and above 75 μm have lower transfer efficiency, that is, are more likely to end up in the recycling. Large particles will have a low charge-to-weight ratio and weaker adhesion to the substrate. Small particles can be difficult to charge and may be difficult to fluidize.

Controlling the particle size distribution during powder production is therefore vital.

6.3.2 FLUIDIZED BED

Powder application by the fluidized bed is very different from the electrostatic spraying. Instead of spraying the powder on the object and subsequent curing in a chamber at high temperature, the object is preheated and dipped into the fluidized bed where the powder that hits the surface melts and forms a film. The fluidized bed consists of a powder container and an air supply, separated by a porous bottom. Clean air passes through the porous bottom in the form of fine bubbles, transforming the powder into a fluidized state, resembling a boiling liquid (Figure 6.3). The powder applied by the fluidized bed has a particle size distribution ranging from 30 to 250 microns, that is, larger particles than applied by electrostatic spraying. The airflow is adjusted so as not to produce dust. Depending on target coating thickness and properties of the powder, the objects are heated to 230°C–450°C and dipped for 2–10 seconds. Typically, coatings on the order of 200–500 μm are produced.

An alternative fluidized bed application technique has been developed, where the preheated object is passed through an electrically charged cloud of powder, which is created above a container of fluidized powder.

Advantages with fluidized bed application are

- The application equipment is rather simple, which means that the investment is small and maintenance costs are low.
- Fast application of rather thick coatings is unique to the fluidized bed method.

The main drawbacks are

- Only high film thickness is achieved.
- Geometry of the object should be such that powder is not trapped.

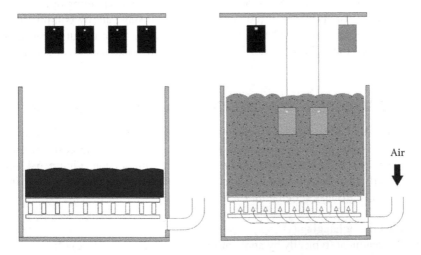

FIGURE 6.3 Application of powder coating in a fluidized bed. Left: Before fluidization. Right: When air is blown into the powder, the volume increases and the objects can be dipped into the bed.

- The object must be preheated, and sometimes postcured in order to get a good film.
- A quite large volume of powder is required to fill the bed. Changing colors can present challenges. If production in multiple colors is desired, investment in one bed per color may be favorable.

6.3.3 FLAME SPRAYING

In flame spraying, the powder only stays in the flame long enough to melt, and is deposited on the object and cooled down before any thermal degradation of the polymer starts. Only thermoplastic coatings can be produced this way because the time at high temperature and liquid state is too short for a curing reaction to complete. The film thickness of the coating is normally much less than the thickness of the substrate, which means that there will be only a moderate temperature increase on the substrate. The temperature of the object will therefore be lower than that for electrostatic or fluidized bed application. Thermal spraying should therefore be regarded as a "cold" process.

The advantage with flame spraying of powder coatings is that there is no size limitation to the object that is coated. Also, any type of substrate can be flame sprayed—metals, wood, polymers, and ceramics. The main drawbacks with the method are moderate adhesion to the substrate, poor protective properties due to a rather open film structure, and a mat surface finish. Due to these limitations, flame spraying of polymer powder coatings is used in limited scale and constitutes the smallest powder volume among the application technologies.

6.3.4 FLOCKING GUN

Application with a flocking gun is done without electrostatic charging of the powder. The powder is simply blown through the gun to the substrate by compressed air. The method is primarily used for preheated substrates, since the dry powder will not adhere well to a cold substrate. The method is much less used than electrostatic spraying.

6.4 ELECTROSTATIC POWDER COATING APPLICATION LINE

One of the great advantages with electrostatic powder coatings is the high level of automation in the application technology. Automation keeps the costs down and contributes to a high and consistent quality in the application. Figure 6.4 shows schematically the various steps in a typical application line. The main steps in the process will be discussed in the following sections:

- Racking or hanging (1)
- Pretreatment (typically by application of a conversion coating) (2)
- Powder application (3)
- Film formation and curing (4)
- Offloading, inspection, and packing (1)

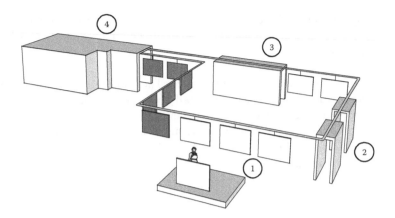

FIGURE 6.4 Powder coating application line.

Only the first and last steps require manual labor, but careful control of the other steps is essential. In small-volume production, all steps can be performed manually.

6.4.1 RACKING OR HANGING

The products to be coated are hung on hooks or racks from a conveyor belt that brings them through the process. These play an important role in grounding the products and will therefore require some attention. The hooks will become powder coated along with the products, which means that the connection to ground will be lost with time, unless the coating is removed from the hooks regularly. The current flowing through the hooks is very low, so some coating is tolerated, but the resistance should be kept below 1 MΩ. However, regular cleaning is essential. The powder can be removed by various methods:

- Thermal cleaning: The binder burns off by pyrolysis, leaving a residue of the inorganic components in the powder on the hooks that must be washed off separately. Alternatively, thermal cleaning can be done in a hot fluidized bed of sand or oxide particles. The hooks are placed in the bed and heated to a temperature where the coating degrades. The grinding effect of the fluidized hard particles removes the degraded coating from the hooks.
- Blast cleaning: The hooks may be blast cleaned by various blasting media, similarly to blast cleaning before painting.
- Chemical cleaning: The hooks are immersed in chemicals or solvents. Thermoplastic coatings can be dissolved, while thermosetting coatings must be degraded chemically. Chemical degradation is usually achieved in heated alkaline baths. The degraded coating either falls off in the bath or is

later sprayed off with water. Very strong solvents may swell thermosetting coatings to such a degree that they can easily be removed mechanically or by washing.

Alternatively, disposable hooks can be used. These are primarily used for small objects, where small hooks are needed. Small hooks are cheap, and it is more cost-efficient to buy new ones than clean the used ones.

Another important aspect of this process step is that the racks are filled as much as possible, without shielding each other from the powder spray. Filling the racks prevents overspraying and excessive recycling of powder, and increases the production rate.

6.4.2 PRETREATMENT

In an automated production line, the pretreatment of the products typically involves application of a conversion coating. The most common conversion coatings are presented briefly in Chapter 10. Most conversion coatings are applied in multistep processes that include degreasing, etching, activation, conversion, and drying, with several intermediate rinsing steps. In the powder application line, the pretreatment is the most complicated, physically largest, and most time-consuming part. The pretreatment is also very critical with respect to the quality of the final coated product [5]. Unsuccessful pretreatment will give a coating with poor adhesion and poor protective properties.

A complicating factor in the pretreatment is that most of the conversion coatings are material specific. They work best on certain materials, and often not at all on others. The most important metals that are powder coated in terms of volume are steel, zinc-coated steel, aluminum, and magnesium. A multimetal pretreatment that works well for products to be exposed in corrosive environments has not been found yet, although there are multimetal treatments that will work in less aggressive environments. Since the pretreatment lines are so specific to the chemistry of the conversion coating, each line will be dedicated to the treatment of one specific metal, at least when it comes to protective powder coatings. This problem can be overcome by performing pretreatment in a separate dipping line. The products to be coated are stacked on racks that are dipped in the baths of the various treatment steps. Many of the treatment steps are common to the various processes, so by adding a few baths, two or more pretreatment processes can be run in the same line. However, dipping lines for pretreatment are usually more expensive to operate than spray lines, since the products have to be treated in two separate production lines, with twice as much manual labor. Job coaters who perform coating for other companies often have dipping lines due to the flexibility, enabling them to coat different metals. Companies that coat their own production tend to choose spraying lines due to the lower cost. Spray lines and dipping lines will give the same quality.

Historically, chromating and phosphating have been the dominant pretreatments prior to powder coating.

6.4.3 POWDER APPLICATION

The powder application technologies were discussed above. The electrostatic spraying guns can be operated manually or by robots, but automated spraying by fixed or oscillating spray guns is more common. The spray guns are mounted in a spray booth, where the conveyor belt carries the products through. An illustration of a powder application booth with recycling is shown in Figure 6.5. The powder that does not attach to the products falls down in the bottom of the spray booth, where it enters the recycling. The recycling uses a fan to pull air into the booth, carrying the powder through the recycling and preventing the overspray powder from escaping the booth. Film thickness is primarily controlled by adjusting the powder flow rate and electrostatic settings in the spray gun, but spray gun distance, movement of the spray gun, and packing density of the racks will also affect the film thickness.

6.4.4 FILM FORMATION AND CURING

After application, the products are taken to the curing oven. Thermoplastic coatings melt and float together into a continuous film, while the air between the particles leaves the film. This is a fairly uncomplicated process that results in good leveling of the coating and uniform film thickness.

For the thermosetting coatings, the curing reaction complicates the process somewhat. Two principles are used for curing of the powder: heat and UV light. Heating is by far the most common, and curing by UV can only be done on powders that are designed for this.

In heat-cured coatings, as the temperature increases, the powder starts melting and floating together. The outer powder melts first, since the substrate has a higher heat capacity and takes longer time to heat. Curing accelerates as the temperature of the powder increases. However, as the curing proceeds, the viscosity of the powder

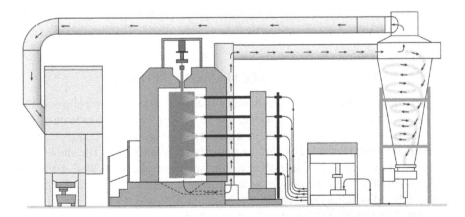

FIGURE 6.5 Powder application booth for automatic spraying, with powder recycling.

increases. All air between the particles must therefore leave the film before the viscosity becomes too high. Curing reactions that produce water molecules must also allow the water to escape the film during the curing. Outgassing at a late stage in the curing process, when the viscosity is high, may result in gas bubbles trapped in the film. If the bubbles leave the film, but the film is too viscous to float together again, pinholes will be formed. Pinholes have a detrimental effect on both decorative and protective properties of the film.

Particle size distribution will also affect the film formation [6]. A certain particle size distribution is beneficial, because it allows for a denser packing of the applied powder, enhances fusion of the particles, and reduces the amount of air between the particles. The irregular shape of the particles will increase the voids in the dry powder film. The dry powder film therefore often has a thickness that is several times higher than the cured film. There is also a relationship between particle size distribution and minimum film thickness. Forming a uniform film with much lower thickness than the particle diameter will be difficult, unless the powder has very good flow-out properties.

UV-cured powder coatings are first heated until the powder particles melt and coalesce into a molten film. The powder may be heated thermally or by infrared light. The film is then cured by irradiation with UV light. The curing only takes seconds, so curing by UV is faster than thermal curing. Forming the film and curing the binder in two steps ensures outgassing before the viscosity of the film starts to increase, which may eliminate the pinhole problem. Since the powder is only heated to melting, and the curing is much faster, the temperature of the substrate will be much lower and elevated for a shorter time, compared with conventional heat-cured coatings. UV curing is therefore particularly suitable for heat-sensitive substrates.

6.4.5 OFFLOADING, INSPECTION, AND PACKING

A major advantage with powder coatings over wet paint is that the product is ready for use immediately after it is offloaded from the production line. The film is cured and already has all the desired properties.

6.5 POWDER COATING OF REBAR AND PIPELINES

Applying fusion-bonded epoxy to concrete rebar and oil and gas pipelines differs from the production line described above, due to the size of the objects to be coated. The process for coating pipe spools is illustrated in Figure 6.6. The pipe is first preheated and blast cleaned. The surface may also be phosphated. The steel pipe is then further heated by induction heating to a temperature of about 230°C, before the powder is applied. Since the powder is applied on a preheated surface, the powder immediately melts, flows together into a continuous film, and starts to cure. The pipe spool may then receive additional coats of thermally insulating materials, like polyethylene or polypropylene. Coating of rebar follows a similar route, but 10 or more rebar rods are coated in parallel.

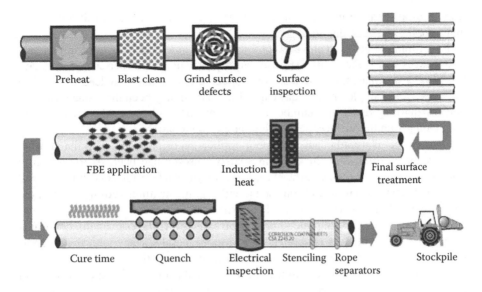

Preheat Blast clean Grind surface Surface
defects inspection

FBE application Induction Final surface
heat treatment

Cure time Quench Electrical Stenciling Rope Stockpile
inspection separators

FIGURE 6.6 Application of FBE on oil and gas pipelines. (Courtesy of Bredero Shaw.)

6.6 COMMON ERRORS, QUALITY CONTROL, AND MAINTENANCE

6.6.1 COMMON ERRORS IN POWDER COATINGS

As with all coatings work, there are a number of typical errors that may occur. A comprehensive overview can be found elsewhere [7]. Some of the most common ones will be briefly discussed here:

- Pinholes
- Low film thickness
- Orange peel
- Poor adhesion

Pinholes in powder coatings are caused by outgassing at a late stage in the film formation process, when curing has proceeded to a stage where viscosity is high and the film has problems floating together again. Anything that will increase the amount of gas that must leave the film will increase the risk for pinholes. A rough or porous substrate will have more air between particles at the substrate. A rough substrate will also be more difficult for the melted powder film to wet. The roughness may be deliberate (blasting) or unwanted, for example, white rust on zinc. If the powder is stored under too-humid conditions, the powder will absorb water that has to leave the film during film formation, increasing the risk for pinholes. Too-high film thickness will also increase the risk for pinholes, since the outer parts of the film may cure before the inner parts have melted completely, trapping gas in the film. Too-fast heating may have a similar effect by rapid increase in film viscosity and trapping of gas in the film.

Low film thickness is a universal problem for organic coatings, for example, over sharp edges and welds. This will also cause problems on powder-coated surfaces. For powder coatings, there are also some specific effects that may lead to low film thickness; for example, the Faraday cage effect has already been mentioned. Hot-dip galvanizing may result in a number of different surface defects, like dross or ash, protruding from the surface, penetrating the powder coating.

Orange peel is usually caused by poor application or problems with the powder. Too-low or too-high film thickness may also cause orange peel. At too-low film thickness, there will not be sufficient powder on the surface for good flow-out and creation of a uniform film, and you end up with a grainy texture. Too-high film thickness may result in a wavy orange peel pattern.

Poor adhesion is usually due to poor pretreatment. If the conversion coating is unsuccessful, the powder is applied on a smooth metal surface with few anchoring points. The surface may also be more chemically active and therefore more susceptible to corrosion. Wet adhesion, that is, adhesion when exposed in a wet environment, will suffer in particular. Incomplete curing will also result in poor adhesion.

6.6.2 QUALITY CONTROL

Quality control of powder coatings is basically done by the same methods as quality control of other organic coatings, for example, adhesion, chalking, and corrosion, which is covered by Chapter 14. Qualicoat has issued a specification of various coating quality tests for powder coatings [8]. A few tests are specific to powder coatings:

- Differential scanning calorimetry (DSC) is used for checking curing of the film and curing properties of the powder. In FBE application lines in particular, DSC is an essential part of the quality control.
- Methyl-ethyl ketone (MEK) test for checking curing: A cotton stick is dipped in MEK and rubbed back and forth in a line a few times on the cured powder coating. If the surface becomes rough and the friction in the movement increases, then the curing is good. Conversely, if the surface is slick and the friction decreases, then the coating has not cured properly.
- Boiling test for checking adhesion: A coated sample is boiled in distilled water for two hours. If the film easily can be peeled off or blisters, then the pretreatment may be inferior.

6.6.3 MAINTENANCE OF POWDER COATINGS

Powder coatings are repaired with wet paint. Similar to repairing wet paint, cleaning the surface is vital for the result. The powder coating should also be rubbed gently, to prepare an anchoring profile to which the repair coating can adhere.

REFERENCES

1. Acmite Market Intelligence. *Global Powder Coating Market*. Ratingen, Germany: Acmite Market Intelligence, 2011.
2. Kehr, J.A. *Fusion-Bonded Epoxy (FBE): A Foundation for Pipeline Corrosion Protection*. Houston: NACE International, 2003.
3. Horinka, P.R., *Powder Coating* 6, 37, 1995.
4. Thies, M.J. Comparison of tribo and corona charging methods: How they work and the advantages of each in powder coating. Presented at *Powder Coating '94*. Taylor Mill, KY: Powder Coating Institute, 1994, pp. 235–251.
5. Bjordal, M.J., et al. *Prog. Org. Coat.* 56, 68, 2006.
6. Mazumder, M.K., et al. *J. Electrostat.* 40–41, 369, 1997.
7. Pietschmann, J. *Powder Coating: Failures and Analyses*. Hannover: Vincentz Network, 2004.
8. Qualicoat. *Specifications for a Quality Label for Liquid and Powder Organic Coatings on Aluminium for Architectural Applications*. Zurich: Qualicoat, 2014.

7 Blast Cleaning and Other Heavy Surface Pretreatments

In broad terms, pretreatment of a metal surface is done for two reasons: to remove unwanted matter and to give the steel a rough surface profile before it is painted. "Unwanted matter" is anything on the surface to be painted except the metal itself and—in the case of repainting—tightly adhering old paint.

For new constructions, matter to be removed is mill scale and contaminants. The most common contaminants are transport oils and salts. Transport oils are beneficial (until you want to paint); salts are sent by an unkind providence to plague us. Transport oil might be applied at the steel mill, for example, to provide a temporary protection to the I-beams for a bridge while they are being hauled on a flatbed truck from the mill to the construction site or the subassembly site. This oil-covered I-beam, unfortunately, acts as a magnet for dust, dirt, diesel soot, and road salts; anything that can be found on a highway will show up on that I-beam when it is time to paint. Even apart from the additional contaminants the oil picks up, the oil itself is a problem for the painter. It prevents the paint from adhering to the steel, in much the same way that oil or butter in a frying pan prevents food from sticking. Pretreatment of new steel before painting is fairly straightforward; washing with an alkali surfactant, rinsing with clean water, and then removing the mill scale with abrasive blasting is the most common approach.

Most maintenance painting jobs mainly consist of repainting existing structures whose coatings have deteriorated. Surface preparation involves removing all loose paint and rust, so that only tightly adhering rust and paint are left. Mechanical pretreatments, such as needle gun and wire brush, can remove loosely bound rust and dirt but do not provide either the cleanliness or the surface profile required for repainting the steel. Conventional dry abrasive blasting is the most commonly used pretreatment; however, wet abrasive blasting and hydrojet cleaning are excellent treatment methods that are also used.

Before any pretreatment is performed, the surface should be washed with an alkali surfactant and rinsed with clean water to remove oils and greases that may have accumulated. Regardless of which pretreatment is used, testing for chlorides (and indeed for all contaminants) is essential after pretreatment and before application of the new paint.

7.1   SURFACE ROUGHNESS

When discussing surface pretreatment, surface roughness is an essential parameter. Roughness is usually measured with a profilometer, that is, a needle that is dragged along a line in the surface, recording the vertical deviation. The profile along the line is then used to calculate various roughness parameters that characterize the profile. The most commonly used parameters are

- Ra: The average deviation from the centerline in the roughness profile.
- Ry: The height difference between the highest peak and the lowest valley along the line.
- Rz: There are actually two slightly different definitions of Rz, given by German and Japanese standards. In the Japanese definition, Rz is defined as the height difference between the average of the five highest peaks and the five lowest valleys. In the German standard, the evaluation length is divided into five equal segments, and Ry is calculated for each segment. Rz is then the average of the five Ry values.

Other parameters that also may be used for characterizing the surface are

- Peak count: The number of peaks per centimeter along the evaluation length
- Tortuosity: The actual length along the roughness profile divided by the nominal distance of the evaluation length
- Wenzel roughness factor: The actual surface area in the surface profile divided by the nominal area

The R parameters will not fully characterize the surface. Figures 7.1 and 7.2 show two cross sections of painted steel. The sample in Figure 7.1 was blast cleaned with grit, while the sample in Figure 7.2 was shot peened. Both surfaces gave approximately the same Ra, Ry, and Rz values, but the surface profiles are very different. The coating on the grit-blasted surface performed well, while the same coating applied on the shot-peened surface failed. It is generally accepted that the sharp profile produced by the grit makes coatings perform better, and grit is therefore usually specified.

FIGURE 7.1   Steel blast cleaned with grit.

FIGURE 7.2 Shot-peened steel.

7.2 INTRODUCTION TO BLAST CLEANING

By far, the most common pretreatment for steel constructions prior to painting is blast cleaning, in which the work surface is bombarded repeatedly with small solid particles. If the individual abrasive particle transfers sufficient kinetic energy to the surface of the steel, it can remove mill scale, rust, or old paint. The kinetic energy (E) of the abrasive particle before impact is defined by its mass (m) and velocity (v), as given in the familiar equation (Equation 7.1)

$$E = (mv^2)/2 \qquad (7.1)$$

Upon impact, this kinetic energy can be used to shatter or deform the abrasive particle, crack or deform old paint, or chip away rust. The behavior of the abrasive, as that of the old coating, depends in part on whether it favors plastic or elastic deformation.

In general, the amount of kinetic energy transferred, and whether it will suffice to remove rust, old paint, and so forth, depends on a combination of

- Velocity and mass of the propelled abrasive particle
- Impact area
- Strength and hardness of the substrate being cleaned
- Strength and hardness of the abrasive particle

In the most commonly used blasting technique—dry abrasive blasting—the velocity of the blasting particles is controlled by the pressure of compressed air. It is more or less a constant for any given dry blasting equipment; the mass of the abrasive particle therefore determines its impact on the steel surface.

In wet abrasive blasting, in which water replaces compressed air as the propellant of the solid blasting media, the velocity of the particles is governed by water pressure. In hydrojet blasting, the water itself is both the propellant and the abrasive (no solid abrasive is used). Both forms of wet blasting offer the possibility to vary the velocity by changing water pressure. It should be noted, however, that wet abrasive blasting is necessarily performed at much lower pressures, and therefore velocities, than hydrojet blasting.

7.3 DRY ABRASIVE BLASTING

Only heavy abrasives can be used in preparing steel surfaces for painting. Lighter abrasive media, such as apricot kernels, plastic particles, glass beads or particles, and walnut shells, are unsuitable for heavy steel constructions. Because of their low densities, they cannot provide the amounts of kinetic energy that must be expended upon the steel's surface to perform useful work. In order to be commercially feasible, an abrasive should be

- Heavy, so that it can bring significant amounts of kinetic energy to the substrate
- Hard, so that it does not shatter into dust or deform plastically (thus wasting the kinetic energy) upon impact
- Inexpensive
- Available in large quantities
- Nontoxic

7.3.1 METALLIC ABRASIVES

Steel is used as abrasive in two forms:

1. Cast as round beads, or shot
2. Crushed and tempered to the desired hardness to form angular steel grit

Scrap or low-quality steel is usually used, often with various additives to ensure consistent quality. Both shot and grit have good efficiency and low breakdown rates.

Steel shot and grit are used for the removal of mill scale, rust, and old paint. This abrasive can be manufactured to specification and offers uniform particle size and hardness. Steel grit and shot can be recycled 100–200 times. Because they generate very little dust, visibility during blasting is superior to that of most other abrasives.

Chilled iron shot or grit can be used for the removal of rust, mill scale, heat treatment scale, and old paint from forged, cast, and rolled steel. This abrasive breaks down gradually against steel substrates, so continual sieving to retain only the large particle sizes may be needed if a rough surface profile is desired in the cleaned surface.

7.3.2 NATURALLY OCCURRING ABRASIVES

Several naturally occurring nonmetallic abrasives are commercially available, including garnet, zircon, novaculite, flint, and the heavy mineral sands magnetite, staurolite, and olivine. However, not all these abrasives can be used to prepare steel for painting. For example, novaculite and flint contain high amounts of free silica, which makes them unsuitable for most blasting applications.

Garnet is a tough, angular blasting medium. It is found in rock deposits in eastern Europe, Australia, and North America. With a hardness of 7–8 Mohr, it is the hardest of the naturally occurring abrasives, and with a specific gravity of 4.1, it is denser than all others in this class except zircon. It has very low particle breakdown on impact, thereby enabling the abrasive to be recycled several times. Among other advantages this confers, the amount of spent abrasive is minimized—an important consideration, for example, when blasting old lead- or cadmium-containing paints. The relatively high cost of garnet limits its use to applications where abrasive can be gathered for recycling. However, for applications where spent abrasive must be treated as hazardous waste, the initial higher cost of garnet is more than paid for by the savings in disposal of spent abrasive.

Nonsilica mineral sands, such as magnetite, staurolite, and olivine, are tough (5–7 Mohr) and fairly dense (2.0–3.0 specific gravity) but are generally of finer particle size than silica sand. These heavy mineral sands—as opposed to silica sand—do not contain free silicates, the cause of the disease silicosis (Section 7.7). In general, the heavy mineral sands are effective for blast cleaning new steel but are not the best choice for maintenance applications [1].

Olivine ($[Mg,Fe]_2[SiO_4]$) has a somewhat lower efficiency than silica sand [2] and occasionally leaves white, chalk-like spots on the blasted surface. It leaves a profile of 2.5 mil or finer, which makes it less suitable for applications where profiling the steel surface is important.

Staurolite is a heavy mineral sand that has low dust levels and, in many cases, can be recycled three or four times. It has been reported to have good feathering and does not embed in the steel surface.

Zircon has higher specific gravity (4.5) than any other abrasive in this class and is very hard (7.5 Mohr). Other good attributes of zircon are its low degree of dusting and its lack of free silica. Its fine size, however, limits its use to specialty applications because it leaves little or no surface profile.

Novaculite is a siliceous rock that can be ground up to make an abrasive. It is the softest abrasive discussed in this class (4 Mohr) and is suitable only for specialty work because it leaves a smooth surface. Novaculite is composed mostly of free silica, so this abrasive is not recommended unless adequate precautions to protect the worker from silicosis can be taken. For the same reason, *flint*, which consists of 90% free silica, is not recommended for maintenance painting.

7.3.3 BY-PRODUCT ABRASIVES

By-product abrasives can be used to remove mill scale on new constructions or rust and old paint in maintenance jobs. These abrasives are made from the residue, or *slag*, left over from smelting metals or burning coal in power plants. Certain melting and boiler slags are glassy, homogeneous mixtures of various oxides with physical properties that make them good abrasives. However, not all industrial slags have the physical properties and nontoxicity needed for abrasives. Boiler (coal), copper, and nickel slags are suitable and dominate this class of abrasives. All three are angular in shape and have a hardness of 7–8 Mohr and a specific gravity of 2.7–3.3; this combination makes for efficient blast cleaning. In addition, none contain significant (1%) amounts of free silica.

TABLE 7.1

Physical Data for By-Product Abrasives

Abrasive	Degree of Dusting	Reuse
Boiler slag	High	Poor
Copper slag	Low	Good
Nickel slag	High	Poor

Source: Modified from Keane, J.D., ed., *Good Painting Practice*, Vol. 1, Steel Structures Painting Council, Pittsburgh, 1982.

Copper slag is a mixture of calcium ferrisilicate and iron orthosilicate. A by-product of the smelting and quenching processes in copper refining, the low material cost and good cutting ability of copper slag make it one of the most economical, expendable abrasives available. It is used in many industries, including major shipyards, oil and gas companies, steel fabricators, tank builders, pressure vessel fabricators, chemical process industries, and offshore yards. Copper slag is suitable for removing mill scale, rust, and old paint. Its efficiency is comparable to that of silica sand [2]. It has a slight tendency to embed in mild steel [3].

Boiler slag—also called *coal slag*—is aluminum silicate. It has a high cutting efficiency and creates a rough surface profile. It too has a slight tendency to embed in mild steel.

Nickel slag, like copper and boiler slag, is hard, sharp, and efficient at cutting, and possesses a slight tendency to embed in mild steel. Nickel slag is sometimes used in wet blasting (see Section 7.4).

It should be noted that because these abrasives are by-products of other industrial processes, their chemical composition and physical properties can vary widely. As a result, technical data reported can also vary widely for this class of abrasives. For example, Bjorgum has reported that copper slag created more blasting debris than nickel slag in trials done in conjunction with repainting of the Älvsborg Bridge in Gothenburg, Sweden [4]. This does not agree with the information reported by Keane [1], which is shown in Table 7.1.

This contradiction in results almost certainly depends on differences in the chemical composition, hardness, and particle size of different sources of the same generic type of by-product abrasive.

Because of the very wide variations possible in chemical composition of these slags, a cautionary note should perhaps be introduced when labeling these abrasives as nontoxic. Depending on the source, the abrasive could contain small amounts of toxic metals. Chemical analyses of copper slag and nickel slag used for the Älvsborg Bridge work have been reported by Bjorgum [4]. Eggen and Steinsmo have also analyzed the composition of various blasting media [5]. The results of both studies are compared in Table 7.2. Comparison of the lead levels in the nickel slags or of the zinc levels in the copper slags clearly indicates that

TABLE 7.2

Levels of Selected Compounds and Elements Found in By-Product Abrasives

Blasting Media	Pb	Co	Cu	Cr	Ni	Zn
Copper slag [5]	0.24%	0.07%	0.14%	0.05%	71 ppm	5.50%
Copper slag [4]	203 ppm	249 ppm	5.6 ppm	1.4 ppm	129 ppm	10 ppm
Nickel slag [5]	73 ppm	0.43%	0.28%	0.14%	0.24%	0.38%
Nickel slag [4]	1.2 ppm	2.3 ppm	4.5 ppm	755 ppm	1.1 ppm	15.6 ppm

Source: Bjorgum, A., Behandling av avfall fra bläserensing, del 3. Oppsummering av utredninger vedrorende behandling av avfall fra blåserensing, Report STF24 A95326, SINTEF, Trondheim, 1995; Eggen, T., and Steinsmo, U., Karakterisering av flater blast med ulike blåsemidler, Report STF24 A94628, SINTEF, Trondheim, 1994.

the amounts of an element or compound can vary dramatically between batches and sources.

By-product abrasives are usually considered one-time abrasives, although there are indications that at least some of them may be recyclable. In the repainting of the Älvsborg Bridge, Bjorgum found that after one use, 80% of the particles were still larger than 250 µm, and concluded that the abrasive could be used between three and five times [4].

7.3.4 MANUFACTURED ABRASIVES

The iron and steel abrasives discussed in Section 7.3.1 are of course man-made. In this section, however, we use the term *manufactured abrasives* to mean those produced for specific physical properties, such as toughness, hardness, and shape. The two abrasives discussed here are very heavy, extremely tough, and quite expensive. Their physical properties allow them to cut very hard metals, such as titanium and stainless steel, and to be recycled many times before significant particle breakdown occurs.

Manufactured abrasives are more costly than by-product slags, usually by an order of magnitude. However, the good mechanical properties of most manufactured abrasives make them particularly adaptable for recycling as many as 20 times. In closed-blasting applications where recycling is designed into the system, these abrasives are economically attractive. Another important use for them is in removing old paints containing lead, cadmium, or chromium. When spent abrasive is contaminated with these hazardous substances, the abrasive might need to be treated and disposed of as a hazardous material. If disposal costs are high, an abrasive that generates a low volume of waste—due to repeated recycling—gains in interest.

Silicon carbide, or *carborundum*, is a dense and extremely hard angular abrasive (specific gravity 3.2, 9 Mohr). It cleans extremely fast and generates a rough surface profile. This abrasive is used for cleaning very hard surfaces. Despite its name, it does not contain free silica.

Aluminum oxide is a very dense and extremely hard angular abrasive (specific gravity 4.0, 8.5–9 Mohr). It provides fast cutting and a good surface profile so that paint can anchor onto steel. This abrasive generates low amounts of dust and can be recycled, which is necessary because it is quite expensive. Aluminum oxide does not contain free silica.

7.4   WET ABRASIVE BLASTING AND HYDROJETTING

In dry abrasive blasting, a solid abrasive is entrained in a stream of compressed air. In *wet abrasive blasting*, water is added to the solid abrasive medium. Another approach is to keep the water but remove the abrasive; this is called *hydrojetting*, or *water jetting*. This pretreatment method depends entirely on water impacting a steel surface at a high enough speed to remove old coatings, rust, and impurities.

The presence of an abrasive medium in the dry or wet pretreatment methods results in a surface with a desirable profile. Hydrojetting, on the other hand, does not increase the surface roughness of the steel. This means that hydrojetting is not suitable for new constructions because the steel will never receive the surface roughness necessary to provide good anchoring of the paint. For repainting or maintenance painting, however, hydrojetting may be used to strip away paint, rust, and so forth, and restore the original surface profile of the steel. It is also used for maintenance in areas where an abrasive can cause damage to other equipment.

Paul [6] mentions that because dust generation is greatly reduced in wet blasting, this method makes feasible the use of some abrasives that would otherwise be health hazards. This should not be taken as an argument to use health-hazardous abrasives, however, because more user-friendly abrasives are available in the market.

7.4.1   TERMINOLOGY

The terminology of wet blasting is confusing, to say the least. The following useful definitions are found in the *Industrial Lead Paint Removal Handbook* [7]:

- *Wet abrasive blast cleaning*: Compressed air propels abrasive against the surface. Water is injected into the abrasive stream either before or after the abrasive exits the nozzle. The abrasive, paint debris, and water are collected for disposal.
- *High-pressure water jetting*: Pressurized water (up to 10,000 psi/700 bar) is directed against the surface to remove the paint. Abrasives are not used.
- *High-pressure water jetting with abrasive injection*: Pressurized water (up to 20,000 psi/1400 bar) is directed against the surface to be cleaned. Abrasive is metered into the water stream to facilitate the removal of rust and mill scale and to improve the efficiency of paint removal. Disposable abrasives are used.

- *Ultra-high-pressure water jetting*: Pressurized water (20,000–40,000 psi/1400–2800 bar; can be higher) is directed against the surface to remove the paint. Abrasives are not used.
- *Ultra-high-pressure water jetting with abrasive injection*: Pressurized water (20,000–40,000 psi/1400–2800 bar; can be greater) is directed against the surface to be cleaned. Abrasive is metered into the water stream to facilitate the removal of rust and mill scale and to improve the efficiency of paint removal. Disposable abrasives are used.

7.4.2 INHIBITORS

An important question in the area of wet blasting is does the flash rust, which can appear on wet-blasted surfaces, have any long-term consequences for the service life of the subsequent painting? A possible preventative for flash rust is adding a corrosion inhibitor to the water.

The literature on rust inhibitors is mixed. Some sources view them as quite effective against corrosion, although they also have some undesirable effects when properly used. Others, however, view rust inhibitors as a definite disadvantage. Which chemicals are suitable inhibitors is also an area of much discussion.

Sharp [8] lists nitrites, amines, and phosphates as common materials used to make inhibitors. He notes problems with each class:

- If run-off water has a low pH (5.5 or less), nitrite-based inhibitors can cause the residue to form a weak but toxic nitrous oxide, which is a safety concern for workers.
- Amine-based inhibitors can lose some of their inhibitive qualities in low-pH environments.
- When using ultrahigh pressure, high temperatures at the nozzle (greater than 140°F [60°C]) can cause some phosphate-based inhibitors to revert to phosphoric acid, resulting in a contaminant buildup.

Van Oeteren [9] lists the following possible inhibitors:

- Sodium nitrite combined with sodium carbonate or sodium phosphate
- Sodium benzoate
- Phosphate, alkali (sodium phosphate or hexametasodium phosphate)
- Phosphoric acid combinations
- Water glass

He also makes the important point that hygroscopic salts under a coating lead to blistering, and therefore, only inhibitors that do not form hygroscopic salts should be used for wet blasting.

McKelvie [10] does not recommend inhibitors for two reasons. First, flash rusting is useful in that it is an indication that salts are still present on the steel surface, and second, he also points out that inhibitor residue on the steel surface can cause blistering.

The entire debate over inhibitor use may be unnecessary. Igetoft [11] points out that the amount of flash rusting of a steel surface depends not only on the presence of water but also very much on the amount of salt present. The implications of his point seem to be this: if wet blasting does a sufficiently good job of removing contaminants from the surface, the fact that the steel is wet afterward does not necessarily mean that it will rust.

Bjørgum found that flash rust will not necessarily decrease coating performance [12]. In a comparison of various surface cleaning methods on rusty steel before painting, ultra-high-pressure water jetting performed better than grit blasting, in spite of flash rust.

7.4.3   ADVANTAGES AND DISADVANTAGES OF WET BLASTING

Wet blasting has both advantages and disadvantages. Some of the advantages are

- More salt is removed with wet blasting (see Section 7.4.4).
- Little or no dust forms. This is advantageous both for protection of personnel and nearby equipment, and because the blasted surface will not be contaminated by dust.
- Precision blasting, or blasting a certain area without affecting nearby areas of the surface, is possible.
- Other work can be done in the vicinity of wet blasting.

Among the disadvantages reported are

- Equipment costs are high.
- Workers have limited vision in and general difficulties in accessing enclosed spaces.
- Cleanup is more difficult.
- Drying is necessary before painting.
- Flash rusting can occur (although this is debatable [see Section 7.4.2]).
- Ultrahigh water jetting is dangerous to the worker. Lethal injuries have been reported.

7.4.4   CHLORIDE REMOVAL

As part of a project testing surface preparation methods for old, rusted steel, Allen [13] examined salt contamination levels before and after treating the panels. Hydrojetting was found to be the most effective method for removing salt, as can be seen in Table 7.3.

The Swedish Corrosion Institute found similar results in a study on pretreating rusted steel [14]. In this study, panels of hot-rolled steel, from which the mill scale had been removed using dry abrasive blasting, were sprayed daily with 3% sodium chloride solution for five months, until the surface was covered with a thick, tightly

TABLE 7.3
Chloride Levels Left after Various Pretreatments

Pretreatment Method	Mean Chloride Concentration (mg/m²)		% Chloride Removal
	Before Pretreatment	After Pretreatment	
Hand wire brush to grade St 3	157.0	152.0	3
Needle gun to grade St 3	116.9	113.5	3
Ultra-high-pressure waterjet to grade DW 2	270.6	17.8	93
Ultra-high-pressure waterjet to grade DW 3	241.9	15.7	94
Dry grit blasting to Sa2½	211.6	33.0	84

Source: Allen, B., *Prot. Coat. Eur.*, 2, 38, 1997.

adhering layer of rust. Panels were then subjected to various pretreatments to remove as much rust as possible and were later tested for chlorides with the Bresle test. Results are given in Table 7.4.

7.4.5 WATER CONTAINMENT

Containment of the water used for pressure washing is an important concern. If used to remove lead-based paint, the water may contain suspended lead particles and needs to be tested for leachable lead using the toxicity characteristic leaching

TABLE 7.4
Chloride Levels after Various Pretreatments

Pretreatment Method	Average Chloride Level (mg/m²)	% Chloride Removal
No pretreatment	349	—
Wire brush to grade SB2	214	39
Needle gun to grade SB2	263	25
Ultra-high-pressure hydrojet, 2500 bar, no inhibitor	10	97
Wet blasting with aluminum silicate abrasive, 300 bar, no inhibitor	16	95
Dry grit blasting to Sa2½ (copper slag)	56	84

Source: Forsgren, A., and Appelgren, C., Comparison of chloride levels remaining on the steel surface after various pretreatments, in *Proceedings of Protective Coatings Europe 2000*, Technology Publishing Company, Pittsburgh, 2000, p. 271.

procedure (see Chapter 8) prior to discharge. Similarly, testing before discharge is needed when using wet blasting or hydrojetting to remove cadmium or chromium pigmented coatings. If small quantities of water are used, it may be acceptable to pond the water until the testing can be conducted [13].

7.5   UNCONVENTIONAL BLASTING METHODS

Dry abrasive blasting will not disappear in the foreseeable future. However, other blasting techniques are currently of interest. Some are briefly described in this section: dry blasting with solid carbon dioxide, dry blasting with an ice abrasive, and wet blasting with soda as an abrasive.

7.5.1   CARBON DIOXIDE

Rice-sized pellets of carbon dioxide (dry ice) are flung with compressed air against a surface to be cleaned. The abrasive sublimes from the solid to the gas phase, leaving only paint debris for disposal. This method reportedly produces lower amounts of dust, and thus containment requirements are reduced. Workers are still exposed to any heavy metals that exist in the paint and must be protected against them.

Disadvantages of this method are its high equipment costs and slow removal of paint. In addition, large amounts of liquid carbon dioxide (i.e., a tanker truck) are needed. Special equipment is needed both for production of the solid carbon dioxide grains and for blasting. Although carbon dioxide is a greenhouse gas, the total amount of carbon dioxide emissions need not increase if a proper source is used. For example, if carbon dioxide produced by a fossil-fuel-burning power station is used, the amount of carbon dioxide emitted to the atmosphere does not increase.

This method can be used to remove paint but is ineffective on mill scale and heavy rust. If the original surface was blast cleaned, the profile is often restored after dry ice blasting. As Trimber [7] sums up, "Carbon dioxide blast cleaning is an excellent concept and may represent trends in removal methods of the future."

7.5.2   ICE PARTICLES

Ice is used for cleaning delicate or fragile substrates, for example, painted plastic composites used in aircrafts. Ice particles are nonabrasive; the paint is removed when the ice causes fractures in the coating upon impact. The ice particles' kinetic energy is transferred to the coating layer and causes conical cracks, more or less perpendicular to the substrate; then lateral and radial cracks develop. When the crack network has developed sufficiently, a bit of coating flakes off. The ice particles then begin cracking the newly exposed paint that was underneath the paint that flaked off. Water from the melted ice rinses the surface free from paint flakes.

Foster and Visaisouk [15] have reported that this technique is good for removing contaminants from crevices in the blasted surfaces. Other advantages are [15]

- Ice is nonabrasive and masking of delicate surfaces is frequently unnecessary.

- No dust results from the breakdown of the blasting media.
- Ice melts to water, which is easily separated from paint debris.
- Ice can be produced on-site if water and electricity are available.
- Escaping ice particles cause much less damage to nearby equipment than abrasive media.

Ice particle blasting has been tested for cleaning of painted compressor and turbine blades on an aircraft motor. The technique successfully removed combustion and corrosion products. The method has also been tested on the removal of hydraulic fluid from aircraft paint (polyurethane topcoat) and removal of a polyurethane topcoat and epoxy primer from an epoxy graphite composite.

7.5.3 SODA

Compressed air or high-pressure water is used to propel abrasive particles of sodium bicarbonate against a surface to be cleaned. Sodium bicarbonate is water soluble; paint chips and lead can be separated from the water and dissolved sodium bicarbonate, thereby reducing the volume of hazardous waste.

The water used with sodium bicarbonate significantly reduces dust. The debris is comprised of paint chips, although it may also be necessary to dispose of the water and dissolved sodium bicarbonate as a hazardous waste unless the lead can be completely removed. The need to capture water can create some difficulties for containment design.

This technique is effective at removing paint but cannot remove mill scale and heavy corrosion. In addition, the quality of the cleaning may not be suitable for some paint systems, unless the surface had been previously blast cleaned. If bare steel is exposed, inhibitors may be necessary to prevent flash rusting.

Most painting contractors are not familiar with this method, but because of similarities to wet abrasive blasting and hydrojetting, they can easily adjust. Because the water mitigates the dust, exposure to airborne lead emissions is significantly reduced but not eliminated; ingestion hazards still exist [15].

7.6 TESTING FOR CONTAMINANTS AFTER BLASTING

Whichever pretreatment method is used, it is necessary before painting to check that the metal surface is free from salts, oils, and dirt.

7.6.1 SOLUBLE SALTS

No matter how good a new coating is, applying it over a chloride-contaminated surface is begging for trouble. Salt contamination can occur from a remarkable number of sources, including road salts if the construction is anywhere near a road or driveway that is salted in the winter. Another major source for constructions in coastal areas is the wind; salt rapidly deposits on every surface in coastal areas. Even the hands of workers preparing the steel for painting contain enough salt to cause blistering after the coating is applied.

Rust in old steel can also be a major source of chlorides. The chlorides that originally caused the rust are caught up in the rust matrix; by their very nature, in fact, chlorides exist at the bottom of corrosion pits—the hardest place to reach when cleaning [16,17].

The ideal test of soluble salts is an apparatus that could be used for nondestructing sampling:

- On-site rather than in the lab
- On all sorts of surfaces (rough, smooth; curved, flat)
- Quickly, because time is money
- Easily, with results that are not open to misinterpretation
- Reliably
- Inexpensively

Such an instrument does not exist. Although no single method combines all these attributes, some do make a very good attempt. All rely upon wetting the surface to leach out chlorides and other salts and then measuring the conductivity of the liquid, or its chloride content, afterward. Perhaps the two most commonly used methods are the Bresle patch and the wetted-filter-paper approach from Elcometer.

The Bresle method is described in the international standard ISO 8502-6. A patch with adhesive around the edges is glued onto the test surface. This patch has a known contact area, usually 1250 mm^2. A known volume of deionized water is injected into the cell. After the water has been in contact with the steel for 10 minutes, it is withdrawn and analyzed for chlorides. There are several choices for analyzing chloride content: titrating on-site with a known test solution; using a conductivity meter; or where facilities permit, using a more sophisticated chloride analyzer. Conductivity meters cannot distinguish between chemical species. If used on heavily rusted steel, the meter cannot distinguish how much of the conductivity is due to chlorides and how much is due simply to ferrous ions in the test water.

The Bresle method is robust; it can be used on very uneven or curved surfaces. The technique is easy to perform, and the equipment inexpensive. Its major drawback is the time it requires; 10 minutes for a test is commonly believed to be too long, and there is a strong desire for something as robust and reliable—but faster.

The filter paper technique is much faster. A piece of filter paper is placed on the surface to be tested, and deionized water is squirted on it until it is saturated. The wet paper is then placed on an instrument (such as the SCM-400 from Elcometer) that measures its resistivity. As in the conductivity measurements discussed above, when this is used for repainting applications, it is not certain how much of the resistivity of the paper is due to chlorides and how much is due simply to rust in the test water. In all, the technique is reliable and simple to implement, although initial equipment costs are rather high.

Neither technique measures all the chlorides present in steel. The Bresle technique is estimated to have around 50% leaching efficiency; the filter paper technique is somewhat higher. One could argue, however, that absolute values are of very limited use; if chlorides are present in any quantity, they will cause problems for the paint. It does not perhaps matter at all that a measurement technique reports 200 mg/m^2, when the correct number was 300 mg/m^2. Both are far too high. There is no general agreement on the tolerance of coatings they tested for salt contamination. However, experimental evidence has shown that chloride salt contamination is more critical than sulfate contamination, and coating systems with a zinc-rich primer have more salt tolerance [18–20]. The effect of salt contamination on coating degradation will also depend on the type of exposure [21]. Axelsen and Knudsen found that small amounts of salt were detrimental to coatings exposed in freshwater. The salt had some effect on coatings in seawater, while little effect was found on coatings in marine atmosphere.

7.6.2 HYDROCARBONS

Like salts, hydrocarbons in the form of oils and grease also come from a variety of sources: diesel fumes, from either passing traffic or stationary equipment motors; lubricating oils from compressors and power tools; grease or oil in contaminated blasting abrasive; oil on operators' hands; and so on. As mentioned above, the presence of oils and grease on the surface to be painted prevents good adhesion.

Testing for hydrocarbons is more complex than is testing for salts for two reasons. First, hydrocarbons are organic, and organic chemistry in general is much more complex than the inorganic chemistry of salts. A simple indicator kit of reagents is quite tricky to develop when organic chemistry is involved. Second, a vast range of hydrocarbons can contaminate a surface, and a test that checks for just a few of them would be fairly useless. What is needed, then, is a test simple enough to be done in the field and powerful enough to detect a broad range of hydrocarbons.

Ever game, scientists have developed a number of approaches for testing for hydrocarbons. One approach is ultraviolet (UV) light, or black lights. Most hydrocarbons show up as an unappetizing yellow or green under a UV lamp. This only works, of course, in the dark, and therefore, testing is done under a black hood, rather like turn-of-the-century photography. Drawbacks are that lint and possibly dust show up as hydrocarbon contaminations. In addition, some oils are not detected by black lights [22]. In general, however, this method is easier to use than other methods.

Other methods that are currently being developed for detecting oils include [23]

- *Iodine with the Bresle patch.* Sampling is performed according to the Bresle method (blister patch and hypodermic), but with different leaching liquids. The test surface is first prepared with an aqueous solution of iodine and then washed with distilled water. Extraction of the dissolved iodine in oil on the surface is thereafter made by the aid of a potassium iodide solution. After extraction of the initially absorbed iodine from the contaminated surface, starch is added to the potassium iodide solution. Assessment of the amount of iodine extracted from the surface is then determined from the degree of blue coloring of the solution. Because the extracted amount of iodine is a measure of the amount of oil residues on the surface, the concentration of the oil on the surface can be determined.
- *Fingerprint tracing method.* Solid sorbent of aluminum oxide powder is spread over the test surface. After heat treatment, the excess of sorbent not strongly attached to the contaminated surface is removed. The amount of

attached sorbent is thereafter scraped off the surface and weighed. This amount of sorbent is a measure of the amount of oil or grease residues on the surface.

- *Sulfuric acid method.* For extraction of oil and grease residues from the surface, a solid sorbent aluminum oxide is used here, too. However, concentrated sulfuric acid is added to the aluminum oxide powder that is scraped off from the contaminated surface. The sulfuric acid solution with the extracted oil and grease residues is then heated. From the coloring of the solution, which varies from colorless to dark brown, the amount of oil and grease residues can be determined.

7.6.3 DUST

Dust comes from the abrasive used in blasting. All blasting abrasives break down to some extent when they impact the surface being cleaned. Larger particles fall to the floor, but the smallest particles form a dust too fine to be seen. These particles are held on the surface by static electricity and, if not removed before painting, prevent the coating from obtaining good adhesion to the substrate.

Examining the surface for dust is straightforward: wipe the surface with a clean cloth. If the cloth comes away dirty, then the surface is too contaminated to be painted. Another method is to apply tape to the surface to be coated. If the tape, when pulled off, has an excessive amount of fine particles attached to the sticky side, then the surface is contaminated by dust. It is a judgment call to say whether a surface is too contaminated because, for all practical purposes, it is impossible to remove all dust after conventional abrasive blasting.

Testing for dust should be done at every step of the paint process because contamination can easily occur after a coating layer has been applied, causing the paint to become tack-free. This would prevent good adhesion of the next coating layer.

Dust can be removed by vacuuming or by blowing the surface down with air. The compressed air used must be clean—compressors are a major source of oil contamination. To check that the compressed air line does not contain oil, hold a clean piece of white paper in front of the airstream. If the paper becomes dirty with oil (or water, or indeed anything else), the air is not clean enough to blow down the surface before painting. Clean the traps and separators and retest until the air is clean and free from water [22].

7.7 DANGEROUS DUST: SILICOSIS AND FREE SILICA

Dry abrasive blasting with silica sand is banned or restricted in many countries because of its link to the disease silicosis, which is caused by breathing excessive quantities of extremely fine particles of silica dust over a long period. This section discusses

- What silicosis is
- What forms of silicon cause silicosis

- Low-free-silica abrasive options
- Hygienic measures to prevent silicosis

7.7.1 WHAT IS SILICOSIS?

Silicosis is a fibronodular lung disease caused by inhaling dust containing crystalline silica. When particles of crystalline silica less than 1 μm are inhaled, they can penetrate deeply into the lungs, through the bronchioles, and down to the alveoli. When deposited on the alveoli, silica causes production of radicals that damage the cell membrane. The alveoli respond with inflammation, which damages more cells. Fibrotic nodules and scarring develop around the silica particles. As the amount of damage becomes significant, the volume of air that can flow through the lungs decreases and, eventually, respiratory failure develops. Epidemiologic studies have established that patients with silicosis are also more vulnerable to tuberculosis; the combination of diseases is called silicotuberculosis and has an increased mortality over silicosis [24–27].

Silicosis has been recognized since 1705, when it was noticed among stonecutters. It has long been recognized as a grave hazard in certain occupations, for example, mining and tunnel boring. The worst known epidemic of silicosis was in the drilling of the Gauley Bridge tunnel in West Virginia in the 1930s. During the construction, an estimated 2000 men were involved in drilling through the rock. Four hundred died of silicosis; of the remaining 1600, almost all developed the disease.

Silicosis is of great concern to abrasive blasters, because the silica breaks down upon impact with the surface being cleaned. The freshly fractured surfaces of silica appear to produce more severe reactions in the lungs than does silica that is not newly fractured [28], probably because the newly split surface of silica is more chemically reactive.

7.7.2 WHAT FORMS OF SILICA CAUSE SILICOSIS?

Not all forms of silica cause silicosis. Silicates are not implicated in the disease, and neither is the element silicon (Si), commonly distributed in the earth's crust and made famous by the semiconductor industry.

Silicates ($-SiO_4$) are a combination of silicon, oxygen, and a metal such as aluminum, magnesium, or lead. Examples are mica, talc, Portland cement, asbestos, and fiberglass.

Silica is silicon and oxygen (SiO_2). It is a chemically inert solid that can be either amorphous or crystalline. *Crystalline silica, also called "free silica," is the form that causes silicosis.* Free silica has several crystalline structures, the most common of which (for industrial purposes) are quartz, tridymite, and cristobalite. Crystalline silica is found in many minerals, such as granite and feldspar, and is a principal component of quartz sand. Although it is chemically inert, it can be a hazardous material and should always be treated with respect.

7.7.3 WHAT IS A LOW-FREE-SILICA ABRASIVE?

A low-free-silica abrasive is one that contains less than 1% free (crystalline) silica. The following are examples of low-free-silica abrasives used in heavy industry:

- Steel or chilled iron, in grit or shot form
- Copper slag
- Boiler slag (aluminum silicate)
- Nickel slag
- Garnet
- Silicon carbide (carborundum)
- Aluminum oxide

7.7.4 WHAT HYGIENIC MEASURES CAN BE TAKEN TO PREVENT SILICOSIS?

The best way to prevent silicosis among abrasive blasters is to use a low-free-silica abrasive. Good alternatives to quartz sand are available (see Section 7.7.3). In many countries where dry blasting with quartz sand is forbidden, these alternatives have proven themselves reliable and economical for many decades.

It is possible to reduce the risks associated with dry abrasive blasting with silica. Efforts needed to do so can be divided into four groups:

- Less toxic abrasive blasting materials
- Engineering controls (such as ventilation) and work practices
- Proper and adequate respiratory protection for workers
- Medical surveillance programs

The National Institute for Occupational Safety and Health (NIOSH) recommends the following measures to reduce crystalline silica exposures in the workplace and prevent silicosis [29]:

- Prohibit silica sand (and other substances containing more than 1% crystalline silica) as an abrasive blasting material and substitute less hazardous materials.
- Conduct air monitoring to measure worker exposures.
- Use containment methods, such as blast cleaning machines and cabinets, to control the hazard and protect adjacent workers from exposure.
- Practice good personal hygiene to avoid unnecessary exposure to silica dust.
- Wear washable or disposable protective clothes at the worksite; shower and change into clean clothes before leaving the worksite to prevent contamination of cars, homes, and other work areas.
- Use respiratory protection when source controls cannot keep silica exposures below the NIOSH recommended exposure limit.
- Provide periodic medical examinations for all workers who may be exposed to crystalline silica.
- Post signs to warn workers about the hazard and inform them about required protective equipment.
- Provide workers with training that includes information about health effects, work practices, and protective equipment for crystalline silica.
- Report all cases of silicosis to state health departments and to the Occupational Safety and Health Administration (OSHA) or the Mine Safety and Health Administration.

For more information, the interested reader is encouraged to obtain the free document "Preventing Silicosis and Deaths from Sandblasting," Department of Health and Human Services Publication (NIOSH) 92-102, which is available at www.osha. gov or by contacting NIOSH at the following address:

Information Dissemination Section
Division of Standards Development and Technology Transfer
NIOSH
4676 Columbia Pkwy
Cincinnati, OH 45226

REFERENCES

1. Keane, J.D., ed. *Good Painting Practice*. Vol. 1. Pittsburgh: Steel Structures Painting Council, 1982.
2. Swedish Corrosion Institute. Handbok i rostskyddsmålning av allmänna stålkonstruktioner. Bulletin 85, 2nd ed. Stockholm: Swedish Corrosion Institute, 1985.
3. NASA (National Aeronautics and Space Administration). Evaluation of copper slag blast media for railcar maintenance. NASA-CR-183744, N90-13681. Washington, DC: NASA, 1989.
4. Bjorgum, A. Behandling av avfall fra bläserensing, del 3. Oppsummering av utredninger vedrorende behandling av avfall fra blåserensing. Report STF24 A95326. Trondheim: SINTEF, 1995.
5. Eggen, T., and U. Steinsmo. Karakterisering av flater blast med ulike blåsemidler. Report STF24 A94628. Trondheim: SINTEF, 1994.
6. Paul, S. *Surface Coatings Science and Technology*. 2nd ed. Chichester: John Wiley & Sons, 1996.
7. Trimber, K.A. *Industrial Lead Paint Removal Handbook*. SSPC 93-02. Pittsburgh: Steel Structures Painting Council, 1993, chaps. 1–9.
8. Sharp, T. *J. Prot. Coat. Linings* 13, 133, 1996.
9. Van Oeteren, K.A. *Korrosionsschutz durch Beschichtungsstoffe*. Munich: Carl-Hanser Verlag, 1980, part 1.
10. McKelvie, A.N. Planning and control of corrosion protection in shipbuilding. Presented at *Proceedings of the 6th International Congress on Metallic Corrosion*, Sydney, 1975, paper 8-7.
11. Igetoft, L. Våtblästring som förbehandling före rostskyddsmålning – litteraturegenomgång. Report 61132:1. Stockholm: Swedish Corrosion Institute, 1983.
12. Bjørgum, A., et al. Repair coating systems for bare steel: Effect of pre-treatment and conditions during application and curing. Presented at CORROSION/2007. Houston: NACE International, 2007, paper 07012.
13. Allen, B. *Prot. Coat. Eur.* 2, 38, 1997.
14. Forsgren, A., and C. Appelgren. Comparison of chloride levels remaining on the steel surface after various pretreatments. In *Proceedings of Protective Coatings Europe 2000*. Pittsburgh: Technology Publishing, 2000, p. 271.
15. Foster, T., and S. Visaisouk. Paint removal and surface cleaning using ice particles. Presented at AGARD SMP Lecture Series on "Environmentally Safe and Effective Processes for Paint Removal," Lisbon, April 27–28, 1995.
16. Mayne, J.E.O. *J. Appl. Chem.* 9, 673, 1959.
17. Appleman, B.R. *J. Prot. Coat. Linings* 4, 68, 1987.
18. Tator, K. B. *J. Protect. Coat. Linings* 27, 50, 2010.
19. Appleman, B. R. *J. Protect. Coat. Linings* 19, 42, 2002.

20. Mitschke, H. *J. Protect. Coat. Linings* 18, 49, 2001.
21. Axelsen, S.B., and O.Ø. Knudsen. The effect of water-soluble salt contamination on coating performance. Presented at CORROSION/2011. Houston: NACE International, 2011, paper 11042.
22. Swain, J.B. *J. Prot. Coat. Linings* 4, 51, 1987.
23. Forsgren, A. *Prot. Coat. Eur.* 5, 64, 2000.
24. Myers, C.E., C. Hayden, and J. Morgan. *Penn. Med.* 60–62, 1973.
25. Sherson, D., and F. Lander. *J. Occup. Med.* 32, 111, 1990.
26. Bailey, W.C., et al. *Am. Rev. Respir. Dis.* 110, 115, 1974.
27. Silicosis and Silicate Disease Committee. *Arch. Pathol. Lab. Med.* 112, 673, 1988.
28. Vallyathan, V., et al. *Am. Rev. Respir. Dis.* 138, 1213, 1988.
29. NIOSH (National Institute of Occupational Safety and Health). NIOSH alert: Request for assistance in preventing silicosis and deaths from sandblasting. Publication (NIOSH) 92-102. Cincinnati, OH: U.S. Department of Health and Human Services, 1992.

8 Abrasive Blasting and Heavy Metal Contamination

In Chapter 7, mention was made of the need to minimize spent abrasive when blasting old coatings containing lead pigments. This chapter covers some commonly used techniques to detect lead, chromium, and cadmium in spent abrasive and methods for disposing of abrasive contaminated with lead-based paint (LBP) chip or dust. Lead receives the most attention, both in this chapter and in the technical literature. This is not surprising because the amount of lead in coatings still in service dwarfs that of cadmium, barium, or chromium.

The growing body of literature on the treatment of lead-contaminated abrasive seldom distinguishes between the various forms of lead found in old coatings, although toxicology literature is careful to do so. Red lead (Pb_3O_4), for example, is the most common lead pigment in old primers, and white lead ($PbCO_3 \cdot Pb[OH]_2$) is more commonly found in old topcoats. It is unknown whether these two lead pigments will leach out at the same rate once they are in landfills. It is also unknown whether they will respond to stabilization or immobilization treatments in a similar manner. A great deal of research remains to be done in this area.

The issue of heavy metals in paint is still depressingly relevant. In an investigation of playgrounds in southwest England in 2015, high levels of lead, chromium, cadmium, and antimony were found in surprisingly many samples [1].

8.1 DETECTING CONTAMINATION

There are really two questions involved in detecting the presence of lead or other heavy metals:

1. Does the old paint being removed contain heavy metals?
2. Will the lead leach out from a landfill?

The amount of a metal present in paint is not necessarily the amount that will leach out when the contaminated blasting media and paint have been placed in a landfill [2–4]. The rate at which a toxic metal leaches out depends on many factors. At first, leaching comes from the surface of the paint particles. The initial rate therefore depends most on the particle size of the pulverized paint. This in turn depends on the condition of the paint to be removed, the type of abrasive used, and the blasting process used [5]. Eventually, as the polymeric backbone of the paint breaks down in a landfill, leaching comes from the bulk of the disintegrating paint particles. The rate at which this happens depends more on the type of resin used in formulating the paint and its chemistry in the environment of the landfill.

8.1.1 Chemical Analysis Techniques for Heavy Metals

Several techniques are available for determining whether toxic metals, such as lead and chromium, exist in paint. Some well-established methods, particularly for lead, are atomic absorption (AA) and inductively coupled plasma atomic emission spectroscopy (ICP-AES). Energy-dispersive x-ray (EDX) in conjunction with scanning electron microscopy (SEM) is a somewhat newer technique.

In the AA and ICP-AES methods, paint chips are dissolved by acid digestion. The amount of heavy metals in the liquid is then measured by AA or ICP-AES analysis. The amount of lead, cadmium, and other heavy metals can be calculated—with a high degree of accuracy—as a total weight percent of the paint. A very powerful advantage of this technique is that it can be used to analyze an entire coating system, without the need to separate and study each layer. Also, because the entire coating layer is dissolved in the acid solution, this method is unaffected by stratification of heavy metals throughout the layer. That is, there is no need to worry about whether the lead is contained mostly in the bulk of the layer, at the coating–metal interface, or at the topmost surface.

EDX-SEM can be used to analyze paint chips quickly. The technique is only semiquantitative: it is very capable of identifying whether the metals of interest are present but is ineffective at determining precisely how much is present. Elements from boron and heavier can be detected. EDX-SEM examines only the surface of a paint chip, to a depth of approximately 5 μm. This is a drawback because the surface usually consists of only binder. It may be possible to use very fine sandpaper to remove the top layer of polymer from the paint; however, this would have to be done very carefully so as not to sand away the entire paint layer. Of course, if the coating has aged a great deal and is chalking, then the topmost polymer layer is already gone. Therefore, analyzing cross sections of paint chips is unnecessary in many cases, particularly for systems with two or more coats. Because coatings are not homogeneous, several measurements should be taken.

8.1.2 Toxicity Characteristic Leaching Procedure

Toxicity characteristic leaching procedure (TCLP) is the method mandated by the U.S. Environmental Protection Agency (EPA) for determining how much toxic material is likely to leach out of solid wastes. A short description of the TCLP method is provided here. For an exact description of the process, the reader should study Method 1311 in EPA Publication SW-846 [6].

In TCLP, a 100 g sample of debris is crushed until the entire sample passes through a 9.5 mm standard sieve. Then 5 g of the crushed sample is taken to determine which extraction fluid will be used. Deionized water is added to the 5 g sample to make 100 ml of solution. The liquid is stirred for 5 minutes. After that time, the pH is measured. The pH determines which extraction fluid will be used in subsequent steps, as shown in Table 8.1. The procedure for making the extraction

TABLE 8.1
pH Measurement to Determine TCLP Extraction Fluid

If the First pH Measurement Is	Then
<5.0	Extraction fluid 1 is used.
>5.0	Acid is added. The solution is heated and then allowed to cool. Once the solution is cooled, the pH is measured again; see below.

If the Second pH Measurement Is	Then
<5.0	Extraction fluid 1 is used.
>5.0	Extraction fluid 2 is used.

TABLE 8.2
Extraction Fluids for TCLP Procedure

	Extraction Fluid 1	Extraction Fluid 2
Step 1	5.7 ml glacial acetic acid is added to 500 ml water.	5.7 ml glacial acetic acid is added to water (water volume < 990 ml).
Step 2	64.3 ml sodium hydroxide is added.	Water is added until the volume is 1 L.
Step 3	Water is added until the volume is 1 L.	
Final pH	4.93 ± 0.05	2.88 ± 0.05

Note: Water used is ASTM D1193 Type II.

fluids is shown in Table 8.2. The debris sample and the extraction fluid are combined and placed in a special holder. The holder is rotated at 30 ± 2 rpm for 18 ± 2 hours. The temperature is maintained at $23 \pm 2°C$ during this time. The liquid is then filtered and analyzed. Analysis for lead and heavy metals is done with AA or ICP-AES.

TCLP is an established procedure, but more knowledge about the chemistry involved in spent abrasive disposal is still needed. Drozdz and colleagues have reported that in the TCLP procedure, the concentrations of lead in basic lead silico chromate are suppressed below the detection limit if zinc potassium chromate is also present. The measured levels of chromium are also suppressed, although not below the detection limit. They attribute this reduction to a reaction between the two pigments that produces a less soluble compound or complex of lead [7].

8.2 MINIMIZING THE VOLUME OF HAZARDOUS DEBRIS

In Chapter 7, we mentioned that choosing an abrasive that could be recycled several times could minimize the amount of spent abrasive. The methods described here attempt to further reduce the amount that must be treated as hazardous debris by separating out heavy metals from the innocuous abrasive and paint binder. The approaches used are

- Physical separation
- Burning off the innocuous parts
- Acid extraction and then precipitation of the metals

At the present time, none of these methods are feasible for the quantities or types of heavy abrasives used in maintenance coatings. They are described here for those wanting a general orientation in the area of lead-contaminated blasting debris.

8.2.1 PHYSICAL SEPARATION

Methods involving physical separation depend on a difference between the physical properties (size and electromagnetics) of the abrasive and those of the paint debris. Sieving requires the abrasive particles to be different in size, and electrostatic separation requires the particles to have a different response to an electric field.

8.2.1.1 Sieving

Tapscott et al. [8] and Jermyn and Wichner [9] have investigated the possibility of separating paint particles from a plastic abrasive by sieving. The plastic abrasive media presumably has vastly different mechanical properties than those of the old paint and, upon impact, is not pulverized in the same way as the coating to be removed.

The boundary used in these studies was 250 µm; material smaller than this was assumed to be hazardous waste (paint dust contaminated with heavy metals). The theory was fine, but the actual execution did not work so well. Photomicrographs showed that many extremely small particles, which the authors believe to be old paint, adhered to large plastic abrasive particles. In this case, sieving failed due to adhesive forces between the small paint particles and the larger abrasive media particles.

A general problem with this technique is the comparative size of the hazardous and nonhazardous particulate. Depending on the abrasive used and the condition of the paint, they may break down into a similar range of particle sizes. In such cases, screening or sieving techniques cannot separate the waste into hazardous and non-hazardous components.

8.2.1.2 Electrostatic Separation

Tapscott et al. [8] have also examined electrostatic separation of spent abrasive. In this process, spent plastic abrasive is injected into a high-voltage, direct current electric field. Material separation depends on the attraction of the particles for the electric field. In theory, metal contaminants can be separated from nonmetal blasting debris. In practice, Tapscott and colleagues reported, the process sometimes produced fractions with heavier metal concentrations, but the separation was insufficient. Neither fraction could be treated as nonhazardous waste. In general, the results were erratic.

8.2.2   LOW-TEMPERATURE ASHING (OXIDIZABLE ABRASIVE ONLY)

Low-temperature ashing (LTA) can be used on oxidizable blasting debris—for example, plastic abrasive—to achieve a high degree of volume reduction in the waste. Trials performed with this technique on plastic abrasive resulted in a 95% reduction in the volume of solid waste. The ash remaining after oxidation must be disposed of as hazardous waste, but the volume is dramatically reduced [10].

LTA involves subjecting the spent abrasive to mild oxidation conditions at moderately elevated temperatures. The process is relatively robust: it does not depend on the mechanical properties of the waste, such as particle size, or on the pigments found in it. It is suitable for abrasives that decompose—with significant solids volume reduction—when subjected to temperatures of 500°C–600°C. Candidate abrasives include plastic media, walnut shells, and wheat starch.

The low-temperature range used in LTA is thought to be more likely to completely contain hazardous components in the solid ash than is incineration at high temperatures. This belief may be unrealistic, however, given that the combustion products of paint debris mixed with plastic or agricultural abrasives are likely to be very complex mixtures [9,10]. Studies of the mixtures generated by LTA of ground walnut shell abrasive identified at least 35 volatile organic compounds (VOCs), including propanol, methyl acetate, several methoxyphenols and other phenols, and a number of benzaldehyde and benzene compounds. In the same studies, LTA of an acrylic abrasive generated VOCs, including alkanols, C_4-dioxane, and esters of methacrylic, alkanoic, pentenoic, and acetic acids [9,10].

LTA cannot be used for mineral or metallic abrasives, which are most commonly used in heavy industrial blasting of steelwork. However, the lighter abrasives required for cleaning aluminum are possible candidates for LTA. Further work would be required to identify the VOCs generated by a particular abrasive medium before the technique could be recommended.

8.2.3   ACID EXTRACTION AND DIGESTION

Acid extraction and digestion is a multistage process that involves extracting metal contaminants from spent blasting debris into an acidic solution, separating the (solid) spent debris from the solution, and then precipitating the metal contaminants as metal salts. After this process, the blasting debris is considered decontaminated and can be deposited in a landfill. The metals in the abrasive debris—now in the precipitate—are still hazardous waste but are of greatly reduced volume.

Trials of this technique were performed by the U.S. Army on spent, contaminated coal slag, mixed plastic, and glass bead abrasives. Various digestive processes and acids were used, and leachable metal concentrations of lead, cadmium, and chromium were measured using the TCLP method before and after the acid digestion.

The results were disappointing: the acid digestion processes removed only a fraction of the total heavy metal contaminants in the abrasives [10]. Based on these results, this technique does not appear to be promising for treating spent abrasive.

8.3 METHODS FOR STABILIZING LEAD

There are concerns about both the permanence and effectiveness of these treatments. The major stabilization methods are explained in this section.

8.3.1 STABILIZATION WITH IRON

Iron (or steel) can stabilize lead in paint debris so that the rate at which it leaches out into water is greatly reduced. Generally, 5%–10% (by weight) of iron or steel abrasive added to a nonferrous abrasive is believed to be sufficient to stabilize most pulverized lead paints [2].

The exact mechanism is unknown, but one reasonable theory holds that the lead dissolves into the leachate water but then immediately plates out onto the steel or iron. The lead ions are reduced to lead metal by reaction with the metallic iron [5], as shown here (Equation 8.1):

$$Pb^{2+} \quad + \quad Fe^0 \quad \rightarrow \quad Pb^0 \quad + \quad Fe^{2+}$$

$$(\text{ion}) \qquad (\text{metal}) \quad (\text{ion}) \qquad (\text{metal}) \qquad (8.1)$$

The lead metal is not soluble in the acetic acid used for extracting metals in the TCLP test (see Section 8.1.2); therefore, the measured soluble lead is reduced. Bernecki et al. [11] make the important point that iron stabilizes only the lead at the exposed surface of the paint chips; the lead inside the paint chip, which comprises most of it, does not have a chance to react with the iron. Therefore, the polymer surrounding the lead pigment may break down over time in the landfill, allowing the bulk lead to leach out. The size of the pulverized paint particles is thus critical in determining how much of the lead is stabilized; small particles mean that a higher percentage of lead will be exposed to the iron.

The permanency of the stabilization is an area of concern when using this technique. Smith [12] has investigated how long the iron stabilizes the lead. The TCLP extraction test was performed repeatedly using paint chips, coal slag abrasive, and 6% steel grit. Initially, the amount of lead leached was 2 mg/L; by the eighth extraction, however, the lead leaching out had increased to above the permitted 5 mg/L. In another series of tests, a debris of spent abrasive and paint particles (with no iron or steel stabilization) had an initial leaching level of 70 mg/L. After steel grit was added, the leachable lead dropped to below 5 mg/L. The debris was stored for six months, with fresh leaching solution periodically added (to simulate landfill conditions). After six months, the amount of lead leached had returned to 70 mg/L. These tests suggest that stabilization of lead with steel or iron is not a long-term solution.

The U.S. EPA has decided that this is not a practical treatment for lead. In an article in the March 2, 1995 issue of the *Federal Register* [13], the issue is addressed by the EPA as follows:

> While it is arguable that iron could form temporary, weak, ionic complexes ... so that when analyzed by the TCLP test the lead appears to have been stabilized, the Agency believes that this "stabilization" is temporary, based upon the nature of the complexing. In fact, a report prepared by the EPA on *Iron Chemistry in Lead-Contaminated Materials* (Feb. 22 1994), which specifically addressed this issue, found that iron-lead bonds are weak, adsorptive surface bonds, and therefore not likely to be permanent. Furthermore, as this iron-rich mixture is exposed to moisture and oxidative conditions over time, interstitial water would likely acidify, which could potentially reverse any temporary stabilization, as well as increase the leachability of the lead.... Therefore, the addition of iron dust or filings to ... waste ... does not appear to provide long-term treatment.

8.3.2 STABILIZATION OF LEAD THROUGH pH ADJUSTMENT

The solubility of many forms of lead depends on the pH of the water or leaching liquid. Hock and colleagues [14] have measured how much lead from white pigment can leach at various pH values using the TCLP test. The results are shown in Figure 8.1.

It is possible to add chemicals, for example, calcium carbonate, to the blasting medium prior to blasting or to the debris afterward, so that the pH of the test solution in the TCLP is altered. At the right pH, circa 9 in Figure 8.1, lead is not soluble in the test solution and thus is not measured. The debris "passes" the test for lead.

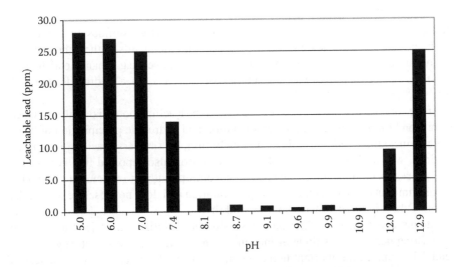

FIGURE 8.1 White lead leachability as a function of pH. (From Hock, V., et al., Demonstration of lead-based paint removal and chemical stabilization using blastox, Technical Report 96/20, U.S. Army Construction Engineering Research Laboratory, Champaign, IL, 1996.)

However, this is not an acceptable technique because the lead itself is not permanently stabilized. The effect, nonsoluble lead, is extremely temporary; after a short time, it leaches precisely as if no treatment had been done [14].

8.3.3 Stabilization of Lead with Calcium Silicate and Other Additives

8.3.3.1 Calcium Silicate

Bhatty [15] has stabilized solutions containing salts of cadmium, chromium, lead, mercury, and zinc with tricalcium silicate. Bhatty proposes that, in water, tricalcium silicate becomes calcium silicate hydrate, which can incorporate in its structure metallic ions of cadmium and other heavy metals.

Komarneni and colleagues [16–18] have suggested that calcium silicates exchange Ca^{2+} in the silicate structure for Pb^{2+}. Their studies have shown that at least 99% of the lead disappears from a solution as a lead–silicate complex precipitate.

Hock and colleagues [14] have suggested a more complex mechanism to explain why cement stabilizes lead: the formation of lead carbonates. When cement is added to water, the carbonates are soluble. Meanwhile, the lead ions become soluble because lead hydroxides and lead oxides dissociate. These lead ions react with the carbonates in the solution and precipitate as lead carbonates, which have limited solubility. Over time, the environment in the concrete changes; the lead carbonates dissolve, and lead ions react with silicate to form an insoluble, complex lead silicate. The authors point out that no concrete evidence supports this mechanism; however, it agrees with lead stabilization data in the literature.

8.3.3.2 Sulfides

Another stabilization technique involves adding reactive sulfides to the debris. Sulfides—for example, sodium sulfide—react with the metals in the debris to form metal sulfides, which have a low solubility (much lower, e.g., than metal hydroxides). Lead, for example, has a solubility of 20 mg/L as a hydroxide, but only 6×10^{-9} mg/L as a sulfide [19].

If the solubility of the metal is reduced, then the leaching potential is also reduced. Robinson [20] has studied sulfide precipitation and hydroxide precipitation of heavy metals, including lead, chromium, and cadmium; he saw less leaching among the sulfides, which also had lower solubility. Robinson also reported that certain sulfide processes could stabilize hexavalent chromium without reducing it to trivalent chromium (but does not call it sulfide precipitation and does not describe the mechanism). Others in the field have not reported this.

Means and colleagues [21] have also studied stabilization of lead and copper in blasting debris with sulfide agents and seen that they could effectively stabilize lead. They make an important point: the mechanical-chemical form of a pulverized paint affects the stabilization. The sulfide agent is required to penetrate the polymer around the metal before it can react with and chemically stabilize the metal. In their research, Means and colleagues used a long mixing time in order to obtain the maximum stabilization effect.

8.4   DEBRIS AS FILLER IN CONCRETE

Solidification of hazardous wastes in Portland cement is an established practice [19]; it was first done in a nuclear waste field in the 1950s [5]. Portland cement has several advantages:

- It is widely available, inexpensive, and of fairly consistent composition everywhere.
- Its setting and hardening properties have been extensively studied.
- It is naturally alkaline, which is important because the toxic metals are less soluble at higher pH levels.
- Leaching of waste in cement has been extensively studied.

Stabilization in Portland cement has one major disadvantage: some of the chemicals found in paint debris have a negative effect on the set and strength development of the cement. Lead, for example, retards the hydration of Portland cement. Aluminum reacts with the cement to produce hydrogen gas, which lowers the strength and increases permeability of the cement [5]. Some interesting work has been done, however, in adding chemicals to the cement to counteract the effects of lead and other toxic metals.

The composition of Portland cement implies that in addition to solidification, stabilization of at least some toxic metals is taking place.

8.4.1   Problems for Concrete Caused by Contaminated Debris

Hydration is the reaction of Portland cement with water. The most important hydration reactions are those of the calcium silicates, which react with water to form calcium silicate hydrate and calcium hydroxide. Calcium silicate hydrate forms a layer on each cement grain. The amount of water present controls the porosity of the concrete: less water results in a denser, stronger matrix, which in turn leads to lower permeability and higher durability and strength [22].

Lead compounds slow the rate of hydration of Portland cement; as little as 0.1% (w/w) lead oxide can delay the setting of cement [23]. Thomas and colleagues [24] have proposed that lead hydroxide precipitates very rapidly onto the cement grains, forming a gelatinous coating. This acts as a diffusion barrier to water, slowing—but not stopping—the rate at which it contacts the cement grains. This model is in agreement with Lieber's observations that the lead does not affect the final compressive strength of the concrete, merely the setting time [23]. Shively and colleagues [25] observed that the addition of wastes containing arsenic, cadmium, chromium, and lead had a delay before setting when mixed with Portland cement, but the wastes' presence had no effect on final compressive strength of the mortar. Leaching of the toxic metals from the cement was greatly reduced compared with leaching from the original (untreated) waste. The same results using cadmium, chromium, and lead were seen by Bishop [26], who proposed that cadmium is adsorbed onto the pore walls of the cement matrix, whereas lead and chromium become insoluble silicates bound into the matrix itself. Many researchers have found that additives, such as sodium silicate, avoid the delayed-set problem; sodium silicate is believed to either form low-solubility metal oxide or silicates, or possibly encapsulate the metal ions in silicate or metal–silicate gel matrices. Either way, the metals are removed from solution before they precipitate on the cement grains.

Compared with lead, cadmium and chromium have negligible effects on the hardening properties of Portland cement [27,28].

8.4.2 Attempts to Stabilize Blasting Debris with Cement

The University of Texas at Austin has done a large amount of research on treatment of spent abrasive media by Portland cement. Garner [29] and Brabrand [30] have studied the effects of concrete mix ingredients, including spent abrasives and counteracting additives, on the mechanical and leaching properties (TCLP) of the resulting concrete. They concluded that it is possible to obtain concrete using spent abrasive with adequate compressive strength, permeability resistance, and leaching resistance. Some of their findings are summarized here:

- The most important factors governing leaching, compressive strength, and permeability were the water–cement ratio and the cement content. In general, as the water–cement ratio decreased and the cement content increased, leaching decreased and compressive strength increased.
- As the contamination level of a mix increased, compressive strength decreased. (It should be noted that this is not in agreement with Shively's [25] results [see Section 8.4.1].)
- Mixes with lower permeability also had lower TCLP leaching concentrations.
- Mixing sequence and time were important for the success of the concrete. The best performance was obtained by thoroughly mixing the dry components prior to adding the liquid components. It was necessary to mix the mortar for a longer period than required for ordinary concrete to ensure adequate homogenization of the waste throughout the mix.
- Set times and strength development became highly unpredictable as the contamination level of the spent abrasives increased.
- Contamination level of the spent abrasives was variable. Possible factors include the condition and type of paint to be removed, the type of abrasive, and the type of blasting process. These factors contribute to the particle size of the pulverized paint and its concentration in the spent blasting abrasives.
- No relationship was found between the leaching of the individual metals and the concrete mix ingredients.

Salt and colleagues [5] have investigated using accelerating additives to counteract the effects of lead and other heavy metals in the spent abrasive on the set, strength, and leaching of mortars made with Portland cement and used abrasive debris. Some of their findings are summarized here:

- Sodium silicate was most effective in reducing the set time of Portland cement mixed with highly contaminated debris, followed by silica fume and calcium chloride. Calcium nitrite was ineffective at reducing the set time for highly contaminated wastes.
- The combination of sodium silicate and silica fume provided higher compressive strength and lower permeability than separate use of these compounds.
- The set time is proportional to the lead–Portland cement ratio; decreasing the ratio decreases the set time. The compressive strength of the mortar, on the other hand, is inversely proportional to this ratio.
- For the most highly contaminated mixtures, accelerators were required to achieve setting.
- All the mortars studied had TCLP leaching concentrations below EPA limits. No correlation between the types and amounts of metals in the wastes and the TCLP leaching results was found.
- The proper accelerator and amounts of accelerator necessary should be determined for each batch of blasting debris (where the batch would be all the debris from a repainting project, which could be thousands of tons in the case of a large structure) by experimenting with small samples of debris, accelerators, and Portland cement.

Webster and colleagues [31] have investigated the long-term stabilization of toxic metals in Portland cement by using sequential acid extraction. They mixed Portland cement with blasting debris contaminated with lead, cadmium, and chromium. The solid mortar was then ground up and subjected to the TCLP leaching test. The solid left after filtering was then mixed with fresh acetic acid and the TCLP test was repeated. This process was done sequentially until the pH of the liquid after leaching and filtering was below 4. Their findings are summarized here:

- The amount of lead leaching was strongly dependent on the pH of the liquid after being mixed with the solid; lead with a pH below 8 began to leach, and the amount of lead leaching rose dramatically with each sequential drop in pH.
- Cadmium also began leaching when the pH of the liquid after being mixed with the solid dropped below 8; the amount of leaching reached a maximum of 6 and then fell off as the pH continued to drop. This could be an artificial maximum, however, because the amount of cadmium was low to begin with; it could be that by the time the pH had dropped to 5, almost all the cadmium in the sample had leached out.
- The authors suggest that the ability of the calcium matrix to resist breakdown (due to acidification) in the concrete is important for the stabilization of lead and cadmium.
- Chromium began leaching with the first extractions (pH 12); the amount leached was constant with each of the sequential extractions until the pH dropped below 6. Because the amount of chromium in the debris was also low, the authors suggest that chromium has no pH dependency for leaching;

instead, it merely leaches until it is gone. This finding was supported by the fact that as the chromium concentration in the blasting debris increased, the TCLP chromium concentration also increased.

- The authors noted that the sequential acid leaching found in their testing was much harsher than concrete is likely to experience in the field; however, it does hint that stabilization of toxic metals with Portland cement will work only as long as the concrete has not broken down.

8.4.3 PROBLEMS WITH ALUMINUM IN CONCRETE

Not all metals can be treated with Portland cement alone; aluminum in particular can be a problem. Khosla and Leming [32] investigated treatment of spent abrasive containing both lead and aluminum by Portland cement. They found that aluminum particles corroded rapidly in the moist, alkaline environment of the concrete, forming hydrogen gas. The gas caused the concrete to expand and become porous, decreasing both its strength and durability. No feasible rapid set (to avoid expansion) or slow set (to allow for corrosion of the aluminum while the concrete was still plastic) was found in this study. (Interestingly, the amount of lead leaching was below the EPA limit despite the poor strength of the concrete.) However, Berke and colleagues [33] found that calcium nitrate was effective at delaying and reducing the corrosion of aluminum in concrete.

8.4.4 TRIALS WITH PORTLAND CEMENT STABILIZATION

In Finland, an on-site trial has been conducted of stabilization of blasting debris with Portland cement. The Koria railroad bridge, approximately 100 m long and 125 years old, was blasted with quartz sand. The initial amount of debris was 150 tons. This debris was run through a negative-pressure cyclone and then sieved to separate the debris into four classes. The amount of "problem debris"—defined in this pilot project as debris containing more than 60 mg of water-soluble heavy metals per kilogram debris—remaining after the separation processes was only 2.5 tons. This was incorporated into the concrete for bottom plates at the local disposal facility [34].

The U.S. Navy has investigated ways to reduce slag abrasive disposal costs in shipyards and found two methods that are both economically and technically feasible: reusing the abrasive and stabilizing spent abrasive in concrete. In this investigation, copper slag abrasive picked up a significant amount of organic contamination (paint residue), making it unsuitable for Portland cement concrete, for which strength is a requirement. It was noted, however, that the contaminated abrasive would be suitable in asphalt concrete [35].

8.4.5 OTHER FILLER USES

Blasting debris can also be incorporated as filler into asphalt and bricks. Very little is reported in the literature about these uses, in particular which chemical forms the heavy metals take, how much leaching occurs, and how permanent the whole arrangement is. In Norway, one company, Per Vestergaard Handelsselkab, has reported sales of spent blasting media for filler in asphalt since 1992 and for filler in brick since 1993. They report that variations in the quality (i.e., contamination levels) of the debris have been a problem [36].

REFERENCES

1. Turner, A., et al. *Sci. Total Environ.* 544, 460, 2016.
2. Trimber, K. *Industrial Lead Paint Removal Handbook.* SSPC Publication 93-02. Pittsburgh: Steel Structures Painting Council, 1993, p. 152.
3. Appleman, B.R. Bridge paint: Removal, containment, and disposal. Report 175. Washington, DC: National Cooperative Highway Research Program, Transportation Research Board, 1992.
4. Harris, J., and J. Fleming. Testing lead paint blast residue to pre-determine waste classification. In *Lead Paint Removal from Industrial Structures.* Pittsburgh: Steel Structures Painting Council, 1989, p. 62.
5. Salt, B., et al. Recycling contaminated spent blasting abrasives in Portland cement mortars using solidification/stabilization technology. Report CTR 0-1315-3F. Springfield, VA: U.S. Department of Commerce, National Technical Information Service (NTIS), 1995.
6. EPA (Environmental Protection Agency). Test methods for evaluating solid waste, physical/chemical methods. 3rd ed., EPA Publication SW-846, GPO doc. 955-001-00000-1. Washington, DC: Government Printing Office, 1993, Appendix II—Method 1311.
7. Drozdz, S., T. Race, and K. Tinklenburg. *J. Prot. Coat. Linings* 17, 41, 2000.
8. Tapscott, R.E., G.A. Blahut, and S.H. Kellogg. Plastic media blasting waste treatments. Report ESL-TR-88-122. Tyndall Air Force Base, FL: Engineering and Service Laboratory, Air Force Engineering and Service Center, 1988.
9. Jermyn, H., and R.P. Wichner. Plastic media blasting (PMB) waste treatment technology. In *Proceedings of the Air and Waste Management Conference.* Vancouver: Air and Waste Management Association, 1991, paper 91-10-18.
10. Boy, J.H., T.D. Rice, and K.A. Reinbold. Investigation of separation, treatment, and recycling options for hazardous paint blast media waste. USACERL Technical Report 96/51. Champaign, IL: Construction Engineering Research Laboratories, U.S. Army Corps of Engineers, 1996.
11. Bernecki, T.F., et al. Issues impacting bridge painting: An overview. FHWA/RD/94/098. Springfield, VA: U.S. Federal Highway Administration, National Technical Information Service, 1995.
12. Smith, L.M. and Tinklenberg, G.L. Lead-containing paint removal, containment, and disposal: Final report. Report No. FHWA-RD-94-100. Federal Highway Administration, U.S Department of Transportation: McLean, VA. 1995.
13. The addition of iron dust to stabilize characteristic hazardous wastes: Potential classification as impermissible dilution. *Federal Register*, vol. 60, no. 41. Washington, DC: National Archive and Record Administration, 1995.
14. Hock, V., et al. Demonstration of lead-based paint removal and chemical stabilization using blastox. Technical Report 96/20. Champaign, IL: U.S. Army Construction Engineering Research Laboratory, 1996.
15. Bhatty, M. Fixation of metallic ions in Portland cement. In *Proceedings of the National Conference on Hazardous Wastes and Hazardous Materials*, Washington, DC, 1987, pp. 140–145.
16. Komarneni, S., D. Roy, and R. Roy. *Cement Concrete Res.* 12, 773, 1982.
17. Komarneni, S. *Nucl. Chem. Waste Manage.* 5, 247, 1985.

18. Komarneni, S., et al. *Cement Concrete Res.* 18, 204, 1988.
19. Conner, J.R. *Chemical Fixation and Solidification of Hazardous Wastes.* New York: Van Nostrand Reinhold, 1990.
20. Robinson, A.K. Sulfide vs. hydroxide precipitation of heavy metals from industrial wastewater. In *Proceedings of the First Annual Conference on Advanced Pollution Control for the Metal Finishing Industry.* Report EPA-600/8-78-010. Washington, DC: Environmental Protection Agency, 1978, p. 59.
21. Means, J., et al. The chemical stabilization of metal-contaminated sandblasting grit: The importance of the physiocochemical form of the metal contaminants in *Engineering Aspects of Metal-Waste Management*, ed. I.K. Iskandar and H.M. Selim. Chelsea, MI: Lewis Publishers, 1992, p. 199.
22. Mindess, S., and J. Young. *Concrete.* Englewood Cliffs, NJ: Prentice-Hall, 1981.
23. Lieber, W. Influence of lead and zinc compounds on the hydration of Portland cement. In *Proceedings of the 5th International Symposium on the Chemistry of Cements*, Tokyo, 1968, vol. 2, p. 444.
24. Thomas, N., D. Jameson, and D. Double. *Cement Concrete Res.* 11, 143, 1981.
25. Shively, W., et al. *J. Water Pollut. Control Fed.* 58, 234, 1986.
26. Bishop, P. *Hazard. Waste Hazard. Mater.* 5, 129, 1988.
27. Tashiro, C., et al. *Cement Concrete Res.* 7, 283, 1977.
28. Cartledge, F., et al. *Environ. Science Technol.* 24, 867, 1990.
29. Garner, A. Solidification/stabilization of contaminated spent blasting media in Portland cement concretes and mortars. Master of science thesis, University of Texas at Austin, TX, 1992.
30. Brabrand, D. Solidification/stabilization of spent blasting abrasives with Portland cement for non-structural concrete purposes. Master of science thesis, University of Texas at Austin, TX, 1992.
31. Webster, M., et al. Solidification/stabilization of used abrasive media for non-structural concrete using Portland cement. NTIS Pb96-111125 (FHWA/TX-95-1315-2). Springfield, VA: U.S. Department of Commerce, National Technical Information Service, 1994.
32. Khosla, N., and M. Leming. Recycling of lead-contaminated blasting sand in construction materials. N. C. Department of Natural Resources & Community Development, June 1988.
33. Berke, N., D. Shen, and K. Sundberg. Comparison of the polarization resistance technique to the macrocell corrosion technique. In *Corrosion Rates of Steel in Concrete.* ASTM STP 1065. Philadelphia: American Society for Testing and Materials, 1990, p. 38.
34. Sörensen, P. *Vanytt* 1, 26, 1996.
35. NSRP (National Shipbuilding Research Program). Feasibility and economics study of the treatment, recycling and disposal of spent abrasives. Final report, Project N1-93-1, NSRP #0529. Charleston, SC: NSRP, ATI Corp., 1999.
36. Bjorgum, A. Behandling av avfall fra bläserensing. Del 3. Oppsummering av utredninger vedrorende behandling av avfall fra blåserensing. Report STF24 A95326. Trondheim: Foundation for Scientific and Industrial Research at the Norwegian Institute of Technology, 1995.

9 Chemical Surface Pretreatments

Chemical pretreatment of metals is typically used before powder coating, coil coating, and other automated coating application processes, because the chemical pretreatment can be implemented in the coating application line. These are also highly automated processes, which means that they have economic benefits and that high production rates are possible.

Chemical pretreatment usually involves application of a conversion coating. The role of the conversion coating is to passivate the surface electrochemically and provide a stable surface layer to which the organic coatings can adhere. Historically, phosphating and chromating have been the dominating conversion coatings. However, due to its high toxicity, chromate conversion coatings are now being replaced by new, more health, safety, and environment (HSE)–friendly processes. This has triggered an enormous amount of research, and many different chemistries have been investigated as potential replacements. Several of these have been developed into commercial processes, replacing chromate. In this chapter, phosphating, chromating, anodizing, and titanium-zirconium (Ti/Zr)–based processes and a process based on Cr(III) are briefly discussed. For the interested reader, here are references to even more processes that are used:

- Cerate based [1,2]
- Vanadate based [3]
- Permanganate based (for magnesium) [4]
- Molybdate based [5]
- Silanes [6]
- Self-assembling molecules [7,8]

Most of the processes that are discussed are found in many variants, but we will not go into details about that. The ambition here is merely to introduce the subject of chemical pretreatment.

9.1 PHOSPHATING

Phosphate conversion coatings are mainly used on steel and zinc, where they are widely used. Phosphate-based processes have also been developed for aluminum, mainly for assembled multimaterial components, but they are less effective there, and other processes are usually preferred when possible. The use of phosphate conversion coatings for corrosion protection started before 1900, and numerous improvements have been introduced since. This is therefore a very mature technology. Narayanan has provided a review of the phosphate conversion coating technology [9].

The phosphate conversion coating does not provide much corrosion protection by itself, and phosphated parts will corrode when exposed to a corrosive environment. However, as a pretreatment before application of an organic coating, it contributes to a very resistant coating system, especially on galvanized and hot-dip galvanized steel.

9.1.1 Formation of the Phosphate Conversion Coating

The metal is treated with a diluted solution of phosphoric acid with one or more metal ions like Zn^{2+}, Mn^{2+}, Ni^{2+}, or Fe^{2+}. When the metal substrate is exposed to this solution, a microgalvanic corrosion process starts, with local anodes and cathodes. On the local cathodes, hydrogen evolution will draw the hydrolytic equilibrium between the various dissociation states of the phosphoric acid toward the insoluble tertiary phosphate:

$$H_3PO_4 = H_2PO_4^- + H^+ = HPO_4^{2-} + 2H^+ = PO_4^{3-} + 3H^+ \tag{9.1}$$

The tertiary phosphate will then precipitate with the metal ions in the solution onto the metal surface, and within a few minutes, the entire surface is covered by the precipitate [10]. Figure 9.1 shows a micrograph of a phosphate conversion coating layer on zinc. After successful phosphating, zinc and steel surfaces will have a matte

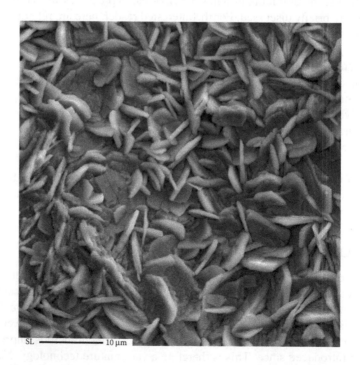

SL ———————— 10 μm

FIGURE 9.1 Phosphate conversion coating on zinc. Picture obtained in electron microscope.

FIGURE 9.2 Steps in a typical phosphating line for galvanized zinc.

and gray appearance, which makes visual quality control quite easy. Gently drawing a fingernail across the phosphated surface leaves a light gray track.

9.1.2 PROCESS STEPS

A typical seven-step phosphating process line for galvanized steel is shown in Figure 9.2. When phosphating steel, an additional acid pickling step is usually needed after the degreasing to remove rust and mill scale, depending on the condition of the material to be treated. The activation step involves treating the surface with a weakly alkaline colloidal dispersion of titanium phosphate. This leads to the formation of crystallites of titanium oxide on the surface, which serve as nucleation points for precipitation of the phosphate conversion coating in the next step. This is particularly important in zinc phosphating, which will produce very coarse crystals of zinc phosphate without the activation.

Different chemistries are used for passivation of the phosphate layer, and as Figure 9.1 indicates, hexavalent chromium was the standard earlier. Passivation with Cr(VI) is now more or less completely replaced with chromate-free alternatives, for example, based on zirconium.

9.1.3 VARIANTS OF PHOSPHATE CONVERSION COATINGS

As mentioned above, various cations can be added to the phosphating bath to modify the properties of the produced conversion coating. Here is a brief overview of the most common variants and their use:

- Iron phosphate: A simple, low-cost process used for pretreatment before coating of steel products to be exposed indoors. Does not provide sufficient corrosion resistance for outdoor exposure.
- Zinc phosphate: Used for pretreatment of steel and galvanized or hot-dip galvanized steel before coating. Zinc phosphating is used as a pretreatment of zinc and steel products to be exposed outdoors.
- Trication phosphate: Contains zinc, manganese, and nickel cations. Was developed for pretreatment of zinc-coated steel before application of the E-coat in the automotive industry (E-coat = electrophoretic application of paint). The substrate is polarized cathodically during E-coating, and alkaline conditions develop on the surface, which the trication phosphating resists [11]. Is now used for pretreatment of galvanized steel before coating in many industries. Suitable for products exposed in corrosive environments.
- Manganese phosphate: Produces a hard, wear-resistant, and somewhat protective phosphate layer. Manganese phosphate is normally not used as a

pretreatment, but a final protection. It is utilized a lot on machined steel parts, like engine components and tools. Virtually all branches of the metal working industry use this process. Posttreatment with oil is common for lubrication and improved corrosion resistance.

9.2 CHROMATE CONVERSION COATINGS: CHROMATING

Chromate conversion coatings were the dominating pretreatment for aluminum, magnesium, and zinc until around 2000. The widespread use was due to their excellent corrosion resistance and coating adhesion. However, since the coating is formed from Cr(VI), which is highly toxic and carcinogenic, it is now being replaced in more and more industries. In the European Union there has, for example, been a ban on the use of Cr(VI) in cars since 2003 and electronics since 2006 [12,13]. More restrictions are expected in the future.

9.2.1 FORMATION OF THE CHROMATE CONVERSION COATING

In the conversion process, Cr(VI) reacts with the metal substrate in a red-ox reaction, where the substrate is oxidized and the Cr(VI) is reduced to Cr(III). Cr(VI) is soluble in the conversion bath, while Cr(III) is not and precipitates as an oxide or hydroxide on the metal surface. On aluminum, the reactions will be

- Oxidation of the substrate:

$$2Al = 2Al^{3+} + 6e^{-} \tag{9.2}$$

- Reduction of chromate:

$$Cr_2O_7^{2-} + 8H^+ + 6e^- = 2Cr(OH)_3 + H_2O \tag{9.3}$$

- Total reaction:

$$2Al + Cr_2O_7^{2-} + 8H^+ = 2Al^{3+} + 2Cr(OH)_3 + H_2O \tag{9.4}$$

Aluminum hydrogen fluoride plays an important role in the process by thinning the natural aluminum oxide surface layer that otherwise would protect the aluminum against the Cr(VI) and slow down the reaction. Kendig and Buchheit have made a thorough review of the chemistry, film formation, toxicity, microstructure, and protection by chromate conversion coatings [14,15].

Figure 9.3 shows a chromate conversion coating on an aluminum surface. The coating has a mud-cracking pattern, which is caused by shrinking of the film during drying. This is normal and does not affect the performance. The grain boundaries in the metal are preferentially etched during the process and are clearly visible in the picture. The pit on the left in the picture was probably caused by a surface-exposed

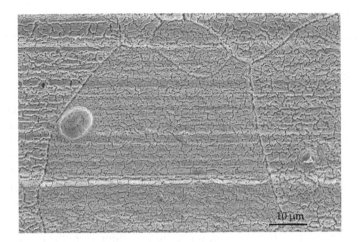

FIGURE 9.3   Chromate conversion coating on aluminum.

noble intermetallic particle. The particle caused local corrosion of the aluminum matrix, until it was undermined and fell out during the alkaline etching. The thickness of chromate conversion coatings is in the order of 100–500 nm, but in the grain boundaries the layer can be absent [16,17]. In addition, the cracks shown in the picture penetrate the layer down to the substrate. In spite of this, the layer is highly protective.

Formation of chromate conversion coatings on magnesium is mainly similar to the formation on aluminum. The details can be found in a review by Gray and Luan [4].

9.2.2   CORROSION PROTECTION

As already mentioned several times, chromate conversion coatings are highly protective against corrosion. The corrosion protection is due to several beneficial properties of the layer, like being hydrophobic and a barrier to water, being stable over a wide pH range, and effectively inhibiting both anodic and cathodic reactions [18]. In addition, it is believed to have a certain capacity for self-healing [19]. During formation of the Cr(III) oxide layer, some Cr(VI) is also precipitated [16,20]. This residue can repair small damages in the conversion coating. The Cr(VI) will react with the aluminum exposed in the damage, forming new protective chromium oxide.

Chromating has successfully been used for corrosion protection of aluminum, magnesium, and zinc, as well as a passivating postrinse on phosphate conversion coatings, as mentioned above. Even the most corrosion-susceptible high-strength aluminum alloys in the 2000 series (alloyed with copper) are protected in very aggressive environments.

9.2.3   CHROMATING PROCESS

The various steps in the chromating process are illustrated in Figure 9.4. The alkaline degreasing will also etch the surface somewhat, in addition to cleaning the surface. In some processes, a stronger alkaline etch is added after the degreasing for

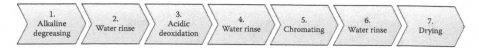

FIGURE 9.4   Process steps in chromating of aluminum.

a deeper etch. The purpose with the acidic deoxidation is to remove smut from the aluminum surface, formed in the alkaline degreasing (and etching). After the chromating, the surface is rinsed and dried, and ready for painting or adhesive bonding.

Another benefit with the chromate conversion coating is that the layer is yellow. An experienced operator can estimate the thickness of the layer by the color nuance, telling whether the conversion was successful or not. This simplifies quality control, compared with most of the chromate-free treatments that produce invisible layers.

Due to the toxicity of the Cr(VI), there are stringent regulations on the treatment of the chemicals, rinsing water, waste, and anything from the process that may contain Cr(VI). This is a hassle and increases the costs of using Cr(VI), and is an additional motivation for changing to chromate-free processes.

As with phosphating, there are different chromating processes, developed for different purposes. For aluminum, yellow, green, and transparent chromating are used:

- Yellow chromating (the process described above): Excellent corrosion protection and adhesion to paint and adhesives.
- Green chromating: The conversion bath contains a mix of chromic and phosphoric acid, producing a green layer of Cr(III) without traces of Cr(VI). Has mainly the same properties as yellow chromating.
- Transparent chromating: A thin conversion coating for applications where the yellow color is unwanted or good electric conductivity is needed. Reduced corrosion protection.

Sheasby and Pinner have given a thorough review of the various chromating processes used on aluminum [21].

9.3   ANODIZING

Anodizing is primarily used as a final treatment of aluminum to obtain corrosion and wear resistance, as well as an appealing visual appearance. However, anodizing has also been used as a pretreatment of aluminum before coating and adhesive bonding for a long time. The aerospace industry in particular has used anodizing for this purpose, but also the architectural industry. GSB International has issued quality requirements for anodizing as pretreatment in the building industry [22]. Also, a special anodizing pretreatment process for coil coating has been developed, which is explained in Section 9.3.1 [21,23–25].

Since anodizing requires immersion of the product in an anodizing bath, implementation in a spray coating or powder coating line is difficult. Pretreatment by anodizing is therefore usually performed in a separate line, which means more handling of the material to be coated. This has given anodizing a reputation for being an expensive pretreatment, and it is less used than the chemical spraying processes. With respect to coating quality, however, anodizing is an excellent process, giving adhesion and corrosion properties comparable to those of chromate conversion coatings [17,26].

9.3.1 DC Anodizing Pretreatment Process

Direct current (DC) anodizing follows the same treatment steps as chromating, as seen by comparing Figure 9.5 with Figure 9.4. For decorative applications, an additional alkaline etching step is often recommended after the degreasing, for example, by GSB International [22]. Anodizing of architectural aluminum is typically performed in 18%–20% sulfuric acid at about 20°C with a current density of 0.8–2.0 A/dm^2 [21,22]. Anodizing for about 10 minutes will produce a 3–8 μm thick layer of aluminum oxide.

The aerospace industry has traditionally anodized in chromic acid, that is, Cr(VI), but has now changed to more HSE-friendly acids. The challenge has been to find an acid that produces an oxide with satisfying corrosion resistance and sufficient adhesive bond strength. Anodizing in a mix of boric and sulfuric acid has been shown to fully reproduce the properties of chromic acid anodizing [27].

Figure 9.6 shows a cross section of the anodizing bath. The aluminum is firmly attached to the rack to secure good electrical contact. The anodizing current flows between the anodic aluminum and the cathodic counterelectrodes. The cathodes can

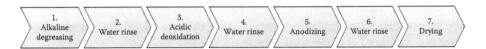

FIGURE 9.5   Process steps in DC anodizing of aluminum.

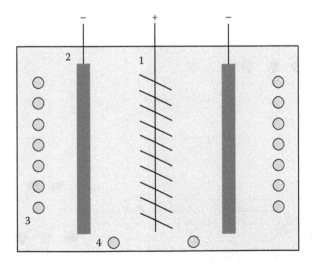

FIGURE 9.6   Anodizing bath seen in cross section. 1, Aluminum on the rack (anode); 2, counterelectrode (cathode); 3, cooling of the bath; 4, air agitation of the bath.

be made of aluminum, typically 6063 or 6101 alloy, lead, stainless steel, or graphite. Previously, lead was usually preferred, but aluminum is now most frequently used. The electrical resistance in the process will result in heating of the bath, and cooling is therefore necessary to maintain a constant temperature. Agitation of the bath is also required to have a homogeneous electrolyte, and is achieved by either low-pressure air bubbling or pumping. There are several excellent reviews about the anodizing process [21,28].

9.3.2 ANODIZING IN COIL COATING

Anodizing in coil coating has to be continuous and very fast, since the metal runs through the process with a speed of about 100 m/minute. The number of process steps also has to be reduced, to keep the physical dimensions of the process within practical limits. This has been achieved by hot alternating current (AC) anodizing. Figure 9.7 shows a simplified illustration of a coil coating line with a hot AC anodizing pretreatment.

The process is typically run in 15%–20% sulfuric acid (or phosphoric acid) at 80°C with an AC density of 20 A/dm^2 (rms). Thus, the anodizing is run at a higher temperature and a higher current density than DC anodizing. The time in the bath is in the order of 5–10 seconds. By using AC, the surface is cleaned in the anodizing bath. In the cathodic current cycle, hydrogen bubbles are formed by the hydrogen reaction, cleaning the surface. The higher current density means faster oxidation of the aluminum. The high temperature reduces the resistance in the cell, enabling the high current density. It also increases the etching of the oxide, so that a porous oxide is formed with good bonding properties to the organic coating.

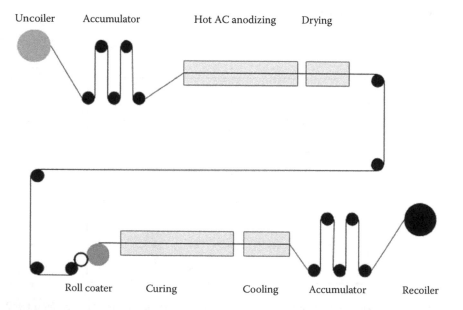

FIGURE 9.7   Coil coating production line.

9.3.3 STRUCTURE AND PROPERTIES OF THE OXIDE LAYER

The anodizing produces a thick layer of aluminum oxide that increases corrosion resistance and coating adhesion. The oxide has a porous structure, as illustrated in Figure 9.8. During anodizing, the oxide grows in the bottom of the oxide, at the interface to the metal. At the same time, it is being etched from the top. The structure of the final oxide depends on the type of acid, temperature, and anodizing time. The pores are formed by acidic etching and are required for transport of reactants to the oxidation site. Anodizing in nonaggressive acids, for example, boric acid or organic acids, will result in a thin, nonporous oxide. In the bottom of the pores is a barrier layer (4) with a thickness of about 20 nm, controlled by applied voltage (of the order 1.4 nm/V). The diameter of the pores (2) is also in the order of 20 nm in sulfuric acid anodizing. In theory, the oxide will form a hexagonal structure as shown to the right, but in practice, it is less organized. Figure 9.9 shows a cross section of the oxide layer in a transmission electron microscope (TEM). The porous structure is clearly shown, although the oxide is somewhat deformed by the sample preparation.

Both DC and hot AC anodized pretreatments are comparable to chromating with respect to corrosion protection and coating adhesion [17,29–32]. DC anodizing can of course produce an oxide that will be highly protective even without an organic coating. The hot AC anodizing results in an oxide thickness of less than 0.5 μm, that is, much thinner than DC anodizing. The main function of the oxide is therefore primarily to secure good coating adhesion. The corrosion protection is then provided by the coating. Thick and protective oxide films can also be formed by AC anodizing at room temperature, but these are not so well suited for decorative applications due to sulfate reduction (during the cathodic half cycle) and incorporation of sulfur in the oxide film.

A thin oxide is also beneficial when making products from the coil-coated material. In many products, the sheet metal is folded 180°, and a thin oxide is then required due to its higher flexibility. Also, softer and less brittle oxide films are formed by anodizing at high temperature.

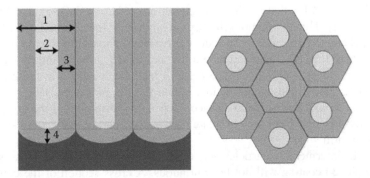

FIGURE 9.8 Schematic illustration of the anodized oxide layer. Cross section to the left and top view to the right. 1, Cell diameter; 2, pore diameter; 3, cell wall thickness; 4, barrier layer thickness.

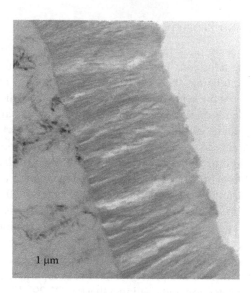

1 µm

FIGURE 9.9   Cross section of anodized oxide in the TEM. The porous structure of the oxide can be seen. The sample preparation has cracked the oxide somewhat.

9.4   TITANIUM- AND ZIRCONIUM-BASED CONVERSION COATINGS

Conversion coatings based on titanium or zirconium or both were the first chromate-free chemical pretreatment processes on the market with a performance that was comparable to that of the chromate conversion coatings.

9.4.1   FORMATION OF THE TITANIUM-ZIRCONIUM LAYER

In the process, the aluminum surface is treated with a weakly acidic solution containing TiF_6^{2-} or ZrF_6^{2-} that results in precipitation of TiO_2 or ZrO_2 on the aluminum surface [33,34]. The aluminum will corrode slightly in the conversion bath due to the low pH and the presence of fluoride. Corrosion of aluminum is a microgalvanic process where the cathodic reaction mainly takes place on noble intermetallic particles containing iron or copper. The pH will increase locally over these particles due to the cathodic hydrogen reaction. When the pH increases, TiO_2 or ZrO_2 is precipitated locally on the metal surface. Hence, the formation of the layer is fundamentally different from the chromate conversion coating that requires a red-ox reaction to form.

Since the intermetallic particles only cover a small fraction of the metal surface, the precipitated coating will not be continuous. A cross section of the conversion coating over an intermetallic particle is shown in Figure 9.10. Over the particle, there is a layer of about 150 nm, while there was little trace of the conversion coating a few micrometers away from the particle. Small particles of TiO_2 or ZrO_2 were found,

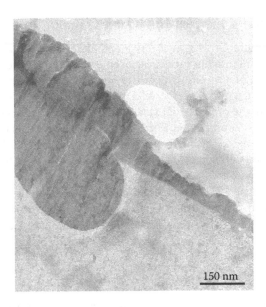

FIGURE 9.10 TiO$_2$-ZrO$_2$ conversion coating on AA6060 aluminum alloy, over an α-Al(Fe,Mn)Si intermetallic particle in the metal surface [33]. (From Lunder, O., et al., *Surf. Coat. Technol.*, 184, 278, 2004.)

though. There are Ti/Zr-based processes that contain organic components, where the organic component will form a continuous layer.

9.4.2 PROCESS AND PROPERTIES

The Ti/Zr conversion coating is applied in the same production line as chromate conversion coatings. Only the solution in the conversion bath is replaced, while the rest of the process is the same. All parts in contact with the conversion bath should be made in AISI 316 stainless steel, though. Since there is little etching of the aluminum in the conversion bath, sufficient etching in the preceding steps is important to obtain the desired quality. This is particularly important for rolled aluminum, where a deformed surface layer can make the coated product highly susceptible to filiform corrosion (see Chapter 12). This layer must be etched away in the pretreatment process, and careful control of the alkaline degreasing and acidic deoxidation steps is therefore important.

Like phosphating on steel and zinc, a Ti/Zr conversion coating improves corrosion protection of coated aluminum by promoting coating adhesion. It provides no corrosion protection on its own.

9.5 CR(III)–BASED CONVERSION COATINGS FOR ALUMINUM

Conversion coatings from Cr(III)–based chemicals have been commercially available since the early 2000s, based on a patent from 2002 [35].

9.5.1 Formation of the Chromium Oxide Coating

The chemicals contain $Cr_2(SO_4)_3$, H_2ZrF_6, and H_2O_2 in a weakly acidic solution. After treatment, the aluminum surface is covered by a dual layer of complex composition with a total thickness of less than 100 nm. The outer layer consists of a complex mix of Cr_2O_3, $Cr_2(SO_4)_3$, ZrO_2, Al_2O_3, and fluorides, while the inner layer is aluminum rich with the presence of oxides and fluorides. The inner layer is assumed to provide the corrosion protection. Formation of the layer seems to go via oxidation of Cr(III) to Cr(VI) by the hydrogen peroxide, so Cr(VI) is actually part of the conversion process, although not added to the chemicals [36]. Nevertheless, the process is accepted as chromate-free.

As shown in Figure 9.11, a continuous layer is formed on the aluminum matrix, but on top of cathodic intermetallic particles, the coating is thicker. The increased precipitation on the particles is probably due to a similar mechanism as for Ti/Zr-based conversion coatings, since the chemicals contain H_2ZrF_6.

9.5.2 Process and Properties

The Cr(III)-based processes can also be applied in the same process line as used for chromating. Only the conversion bath is replaced.

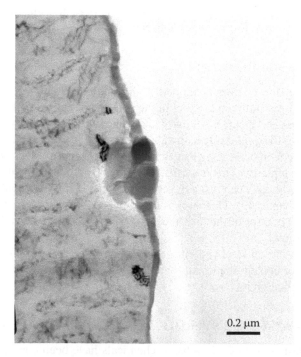

0.2 µm

FIGURE 9.11   Conversion coating on aluminum formed from a Cr(III)–based process.

As a pretreatment before coating, the Cr(III)-based process gives excellent adhesion and filiform corrosion resistance [37]. These coatings seem to be able to fully replace the chromate conversion coatings, since they also provide good corrosion protection of aluminum without the organic coating. They are so far the only chromate-free conversion coatings that meet the performance requirements given by the MIL-DTL-81706 standard [38].

REFERENCES

1. Hughes, A.E., et al. *Surf. Coat. Technol.* 203, 2927, 2009.
2. Harvey, T.G. *Corros. Eng. Sci. Technol.* 48, 248, 2013.
3. Iannuzzi, M., and G.S. Frankel. *Corros. Sci.* 49, 2371, 2007.
4. Gray, J.E., and B. Luan. *J. Alloys Compounds* 336, 88, 2002.
5. Yong, Z., et al. *Appl. Surf. Sci.* 255, 1672, 2008.
6. De Graeve, I., et al. 1—Silane films for the pre-treatment of aluminium: Film formation mechanism and curing. In *Innovative Pre-Treatment Techniques to Prevent Corrosion of Metallic Surfaces*. Cambridge, UK: Woodhead Publishing, 2007, p. 1
7. Terryn, H. *ATB Metallurgie* 38, 41, 1998.
8. Schreiber, F. *Prog. Surf. Sci.* 65, 151, 2000.
9. Narayanan, T.S.N.S. *Rev. Adv. Mater. Sci.* 9, 130, 2005.
10. Utgikar, V., et al. *Biotechnol. Bioeng.* 82, 306, 2003.
11. Debnath, N.C., and G.N. Bhar. *Eur. Coat. J.* 17, 46, 2002.
12. European Union. End-of-life vehicles. Directive 2000/53/EC, 2000.
13. European Union. Restriction of the use of certain hazardous substances in electrical and electronic equipment. Directive 2002/95/EC, 2002.
14. Kendig, M.W., and R.G. Buchheit. *CORROSION* 59, 379, 2003.
15. Frankel, G.S., and R.L. McCreery. *Interface* 10, 34, 2001.
16. Lunder, O., et al. *Corros. Sci.* 47, 1604, 2005.
17. Knudsen, O.Ø., et al. *Corros. Sci.* 46, 2081, 2004.
18. Kendig, M., et al. *Surf. Coat. Technol.* 140, 58, 2001.
19. Kendig, M.W., et al. *Corros. Sci.* 34, 41, 1993.
20. Ayllon, E.S., and B.M. Rosales. *CORROSION* 50, 571, 1994.
21. Sheasby, P.G., and R. Pinner. *The Surface Treatment and Finishing of Aluminium and Its Alloys*. 6th ed. Materials Park, OH: ASM International, 2001.
22. GSB AL 631. International quality regulations for the coating of building components. Schwäbisch Gmünd, Germany: GSB International, 2013.
23. Lister, L. Production of laquered aluminium or aluminium alloy strip or sheet. Patent BP 1,235,661, 1971.
24. Davies, N. C., and P. G. Sheasby, Anodic aluminium oxide film and method of forming it, Patent no. US 4681668 A, 1987.
25. Wootton, E.A. The pretreatment and lacquering of aluminum strip for packaging applications. *Sheet Metal Ind.* 53, 297, 1976.
26. Knudsen, O.Ø., et al. *ATB Metallurgie* 43, 175, 2003.
27. Critchlow, G.W., et al. *Int. J. Adhesion Adhesives* 26, 419, 2006.
28. Grubbs, C.A. Anodizing of aluminum. In *Metal Finishing*, ed. R.E. Tucker. New York: Metal Finishing Magazine, 2011, p. 401.
29. Lunder, O., et al. *Int. J. Adhesion Adhesives* 22, 143, 2002.
30. Bjørgum, A., et al. *Int. J. Adhesion Adhesives* 23, 401, 2003.
31. Johnsen, B.B., et al. *Int. J. Adhesion Adhesives* 24, 153, 2004.
32. Johnsen, B.B., et al. *Int. J. Adhesion Adhesives* 24, 183, 2004.

33. Lunder, O., et al. *Surf. Coat. Technol.* 184, 278, 2004.
34. Nordlien, J.H., et al. *Surf. Coat. Technol.* 153, 72, 2002.
35. Matzdorf, C., et al. Corrosion resistant coatings for aluminum and aluminum alloys. Patent US 6,375,726, 2002.
36. Qi, J.T., et al. *Surf. Coat. Technol.* 280, 317, 2015.
37. Knudsen, O.Ø., et al. *ATB Metallurgie* 45, 2006.
38. MIL, -DTL-81706B Chemical conversion materials for coating aluminum and aluminum alloys. Lakehurst, NJ: Naval Air Warfare Center Aircraft Division, 2002.

10 Adhesion and Barrier Properties of Protective Coatings

A wide range of coating properties have to be optimized, often with compromises between them, for any coating product, for example, application, curing, visual, and protective properties, to name a few high-level properties. The protective properties are very dependent on adhesion and barrier properties, and these two properties therefore receive special attention in this chapter.

10.1 ADHESION

Adhesion is a premise for protective coatings. Without adhesion, we do not have a coating. On the other hand, a firm correlation between adhesion strength and corrosion protection has not been documented [1]. This may partly be explained by the fact that there are no good methods available for quantifying adhesion, at least for state-of-the-art protective coatings. The most commonly used method for adhesion measurement, the pull-off test (ISO 4624, ASTM D4541), usually measures coating cohesion, since most coatings fail cohesively in the test. The cross-cut test, which is another commonly used method, is a qualitative method (ISO 2409, ASTM D3359). Studies of model coatings with low adhesion may give quantitative results about adhesion, but will not necessarily be relevant for strongly adhering state-of-the-art protective coatings. Adhesion measurement is discussed in Chapter 15. The role of adhesion is to create the necessary conditions so that corrosion protection mechanisms can work. A coating cannot passivate the metal surface or create a path of extremely high electrical resistance at the metal surface unless it is in intimate contact—at the atomic level—with the surface. The more chemical and physical bonds there are between the surface and coating, the closer the contact and the stronger the adhesion. An irreverent view could be that the higher the number of sites on the metal that are taken up in bonding with the coating, the lower the number of sites remaining available for electrochemical mischief. Or as Koehler expressed it,

> The position taken here is that from a corrosion standpoint, the degree of adhesion is in itself not important. It is only important that some degree of adhesion to the metal substrate be maintained. Naturally, if some external agency causes detachment of the organic coating and there is a concurrent break in the organic coating, the coating will no longer serve its function over the affected area. Typically, however, the detachment occurring is the result of the corrosion processes and is not quantitatively related to adhesion. [2]

Adhesion depends not only on the coating chemistry, but also on the nature of the substrate, surface pretreatment, and the environment to which the coating is exposed. Loss of adhesion due to water, corrosion products, or other factors has been subject to much research. In summary, good adhesion of the coating to the substrate could be described as a "necessary but not sufficient" condition for good corrosion protection. For all the protection mechanisms described earlier, good adhesion of the coating to the metal is a necessary condition. However, good adhesion alone is not enough; adhesion tests in isolation cannot predict the ability of a coating to control corrosion [3].

10.1.1 ADHESION FORCES

Mittal [4] and Lee [5] have written reviews about the nature of adhesion. The most important forces in polymer-to-metal adhesion can be classified into the following four groups: chemical bonds, intermolecular forces, molecular interactions, and mechanical interlocking. Except for the mechanical interlocking, these bonds are of a chemical nature. Each group is briefly described below. However, the importance of the various forces for the total adhesion is a matter of some discussion.

Chemical bonds, that is, covalent or ionic bonds, are the strongest bonds between atoms, but there is some doubt about to what degree they exist as an adhesive bond. These bonds work at very short range, on the order of 0.1 nm, and the bond energy is usually between 200 and 800 kJ/mol. Construction materials such as steel and aluminum are covered by an oxide or hydroxide layer when they are painted, and bonds must be formed to this layer. However, binders in protective organic coatings do not contain functional groups that are capable of forming covalent bonds to metal oxides and hydroxides. Besides, covalent bonds between organic substances and metal atoms are in most cases easily hydrolyzed by water [6]. The addition of organosilanes to paint is known to increase adhesion, and may work by creating covalent bonds. The organic tail of the silane may bond to the binder, and the siloxane head may bond to the metal surface oxide [7].

According to Lee, the most important adhesion forces between metals and polymers are the intermolecular forces or Van der Waals forces [5]. This is a group of columbic forces that act over some distance (>10 nm). The group includes dispersion forces, dipole–dipole attractions, and induction forces. Dispersion forces are attractions between temporary dipoles in nonpolar molecules. Dipole–dipole attractions are forces between permanent dipoles, while induction forces are forces between a permanent dipole and an induced dipole. The bond strength of a dipole–dipole bond may be as high as 20 kJ/mol, while dispersion forces are much weaker.

Molecular interactions are in an intermediate position between the chemical bonding and the intermolecular forces [5]. They are also short-range forces, working below 0.3 nm. Bond strength is higher than for the Van der Waals forces, but lower than for the chemical bonds. These forces are difficult to explain in a simple manner, but can generally be described as interactions between electron donors and electron acceptors. The highest occupied molecular orbital (HOMO) of the electron donor interacts with the lowest unoccupied molecular orbital (LUMO) of the electron acceptor. The Lewis acid–base theory can also be used to describe these adhesion forces [8].

Mechanical interlocking is caused by geometric irregularities on the surface of the substrate. The coating may cling to the irregularities in the surface by crimping around peaks or anchoring beneath them. Mechanical bonds have been claimed to be of minor importance for adhesion, unless the substrate is porous [9]. However, blast cleaning will produce a profile with features beneath which the coating can anchor. A study of coatings on machined surfaces showed that skew peaks in the machining profile where the coating could hook underneath had a huge impact on adhesion and protective properties of a commercial epoxy coating [10]. Mechanical interlocking therefore seems to be a very important adhesion mechanism.

The chemical adhesion forces will depend on the chemical nature of the binder and the substrate. Nonpolar binders such as polyolefins and fluoropolymers have weak adhesion because most of the chemical adhesion forces depend on functional groups in the binder. Without functional groups such as hydroxyl, carboxyl, and amines in the binder, its ability to form chemical adhesion forces is limited. On the other hand, adhesion of, for example, epoxy coatings to smooth surfaces is also weak [11], indicating that both chemical and mechanical forces contribute to adhesion.

10.1.2 Effect of Surface Roughness on Adhesion

The importance of surface roughness on coating adhesion is well documented through both practical experience and laboratory investigations [12–14]. However, as mentioned above, the lack of methods for quantifying coating adhesion impedes the investigation of the correlation between surface roughness and adhesion. In the context of protective coatings, surface roughness is usually achieved by blast cleaning (Chapter 8), or other mechanical surface treatments, such as grinding or steel brushing. The effect of surface roughness on adhesion may be due to both increased contact area between the coating and the substrate, and mechanical interlocking.

Mechanical adhesion forces are caused by geometric irregularities on the surface of the substrate. The coating may cling to the irregularities in the surface by crimping around peaks or anchor beneath them. According to van der Leeden and Frens, not all rough surfaces possess the ability for mechanical interlocking [15]. In Figure 10.1a, only the irregularity marked by arrows will give a hook and anchor effect, while any effect on mechanical adhesion from the rest of the peaks will be from crimping forces in the coating. However, all the irregularities will increase resistance to lateral forces. Internal stress in the paint film will result in lateral forces, and epoxy coatings with lower internal stresses showed better adhesion properties and resistance against corrosion in cyclic tests [16]. Impacts will also result in lateral forces, due to the sideways squeezing of the coating under the impinging object. Coatings on smooth surfaces will have poor impact toughness. Increased contact area between the coating and the substrate will increase the chemical adhesion by increasing the number of bonds, unless air is trapped in the roughness and the coating is unable to wet the surface. This situation is illustrated in Figure 10.1b. Incomplete wetting decreases the contact area between the coating and the substrate. The voids may also serve as initiation points for corrosion.

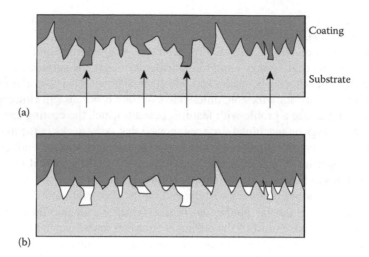

FIGURE 10.1   Various shapes of irregularities on a surface that will affect adhesion. (a) Only the sites indicated by arrows will achieve a hook and anchor effect. (b) Incomplete wetting of the substrate will reduce adhesion, and the voids may serve as initiation points for corrosion.

A profile parameter was proposed in the 1930s by Wenzel, who derived the ratio of the contact area divided by the projected geometric area, known as Wenzel's roughness factor, r [17]. Correlation between the Wenzel roughness factor and adhesion has been found, but it is difficult to distinguish between the effect of the increased contact area and mechanical adhesion forces. Similar to the Wenzel roughness factor, but for a cross-sectional line instead of an area, tortuosity is defined as the ratio between the actual microscopic interface length of the surface profile and the apparent macroscopic length profile [18]. Sørensen found a direct correlation between cathodic disbonding and the tortuosity of the steel substrate.

10.1.3   EFFECT OF SURFACE CHEMISTRY ON ADHESION

Modifying the surface chemistry of the substrate is another way of improving adhesion. On metals, this is typically achieved by applying conversion coatings. The conversion coating improves adhesion by either introducing microscopic roughness on the surface, introducing new chemistry on the surface, removing impurities, or stabilizing the surface chemically, or a combination of these. Frequently used conversion coatings include phosphating, chromating, anodizing, titanium–zirconium-based, and chromium III–based conversion coatings. The various types of conversion coatings are discussed in Chapter 9.

Again, it is difficult to distinguish between the various mechanisms by which the conversion coating improves adhesion. However, there can be no doubt about the cleaning effect. Adhesion is always adversely affected by impurities or unwanted phases at the interface. The effect is elaborated by Bikerman in the weak boundary layer theory [19]. The coating may adhere to surface impurities, but the impurities may not adhere to the substrate, serving as a weak boundary layer and causing adhesion loss. Cleaning is vital and probably the single error responsible for most coating failure incidents.

10.1.4 WET ADHESION

A coating can be saturated with water, but if it adheres tightly to the metal, it can still prevent sufficient amounts of electrolytes from collecting at the metal surface for the initiation of corrosion. How well the coating clings to the substrate when it is saturated is known as *wet adhesion*. Adhesion under dry conditions is probably overrated; wet adhesion, on the other hand, is crucial to corrosion protection. It should be noted that wet adhesion is a coating property and not a failure mechanism. Permanent adhesion loss due to humid or wet circumstances also exists and is called *water disbondment*.

Coatings with good dry adhesion may have poor wet adhesion [20]. The same polar groups on the binder molecules that create good dry adhesion can wreak mischief by decreasing water resistance at the coating–metal interface—that is, they decrease wet adhesion [21]. Another important difference is that dry adhesion, once lost, cannot be recovered. Loss of adhesion in wet conditions, on the other hand, can be reversible. The original dry adhesion strength may not necessarily be obtained [22,23]. However, after cyclic corrosion testing it is frequently found that adhesion after the test is higher than for unexposed samples [24]. The increased adhesion may be due to additional curing in the high-temperature exposure in the cyclic test, but the results definitely show that strong adhesion is retained after wet exposure. Recovery of adhesion strength after wet exposure has also been found for epoxy adhesives [25].

Stratmann et al. have studied water at the metal–polymer interface by reflectance Fourier transform infrared (FTIR) spectroscopy [26]. The coating they studied was a 100 µm thick unpigmented alkyd resin that was put on a ZnSe prism coated with 10 nm Fe. They found that the concentration of water at the polymer–substrate interface was 1.1 mol/L, which was only slightly higher than the concentration of water found in the bulk polymer (0.8 mol/L). The formation of a separate aqueous phase at the interface was therefore excluded. The chemical state of water at the interface was studied by comparing the FTIR spectrum of a coating before and after exposure. When the coating absorbed water, the carbonyl absorption band shifted, so they concluded that the water was associated with the carbonyl groups in the alkyd.

It is reasonable to assume that water molecules at the metal–coating interface will affect the chemical adhesion forces described above, due to its polar nature. Transformation of surface metal oxides to hydroxides during wet exposure will likely also affect adhesion [27]. It has been suggested that chemical adhesion could account for reduced adhesion in wet circumstances, whereas mechanical locking may account for residual wet adhesion [23].

10.1.5 IMPORTANT ASPECTS OF ADHESION

Two aspects of adhesion are important: the initial strength of the coating–substrate bond and what happens to this bond as the coating ages.

A great deal of work has been done to develop better methods for measuring the initial strength of the coating–substrate bond. Unfortunately, the emphasis on measuring initial adhesion may miss the point completely. It is certainly true that good adhesion between the metal and the coating is necessary for preventing corrosion under the coating. However, it is possible to pay too much attention to measuring the difference between good initial adhesion and excellent initial adhesion, completely missing the question of whether that adhesion is maintained. In other words, as long as the coating has good initial adhesion, then it does not matter whether that adhesion is simply good or excellent. What matters is what happens to the adhesion during exposure and over time. This aspect is much more crucial to coating success or failure than is the exact value of the initial adhesion.

Adhesion tests on wet and aged coatings are useful not only to ascertain if the coatings still adhere to the metal, but also to yield information about the mechanisms of coating failure. This area deserves greater attention, because studying changes in the failure loci in adhesion tests before and after weathering can yield a great deal of information about coating deterioration.

10.2   BARRIER PROPERTIES

The most important corrosion protection mechanism of organic coatings is to prevent the formation of a water phase at the metal surface. However, in order to have durable protection, the coating must be a barrier to ions. The ionic resistance of the coating reduces the flow of current available for anode–cathode corrosion reactions. In other words, water—but not ions—may readily permeate most coatings.

10.2.1   DIFFUSION IN POLYMERS

Diffusion in polymers has been investigated for more than a century. Several reviews about the topic exist. Two classic books in the field are

- *Diffusion in Polymers* by Crank and Park [28]
- *Transport in Polymer Films* by Park [29]

Fick's first law gives the kinetics for ideal diffusion of a species in a stationary concentration gradient (Equation 10.1):

$$J = -D\frac{dC}{dx} \tag{10.1}$$

J is the flux of the diffusing species, D is the diffusion coefficient, and dC/dx is the concentration gradient. The diffusion coefficient D describes how easily a species moves in a particular polymer. The equation can be used to calculate how much of the diffusing species passes through the film per area unit. For a protective paint film, the concentration gradient dC/dx is across the film. The concentration gradient is the driving force for the diffusion.

Often, the concentration gradient in the polymer is unknown. Then it is more practical to express the transport of a species through a coating by the permeability coefficient (Equation 10.2):

$$P = D \times S \tag{10.2}$$

P is the permeability coefficient, D is the diffusion coefficient in Fick's first law, and S is the partition coefficient of the diffusing species between the polymer and the outside medium, for example, an electrolyte ($S = C_{polymer}/C_{electrolyte}$). The flux can then be expressed by the difference in concentration of the diffusing species in the electrolytes on both sides of the polymer (Equation 10.3):

$$J = -P \frac{\Delta C_{electrolyte}}{\delta} \tag{10.3}$$

Fick's second law gives the kinetics for ideal diffusion of a species in a changing concentration gradient (Equation 10.4):

$$\frac{\partial C}{\partial t} = D \frac{\partial^2 C}{\partial x^2} \tag{10.4}$$

This equation is useful when studying transients. With proper boundary conditions, Equation 10.4 can be integrated to equations describing, for example, the desorption of water from a paint film or time lag for diffusion of oxygen through a paint film.

To get a conceptual model for diffusion of nonionic molecules in polymers, the free volume theory is helpful [30]. The free volume in a polymer is the volume not occupied by the polymer chains, and thereby is accessible for the permeating species. In principle, the free volume is defined for a polymer with no pigments or voids. The free volume for a given polymer is mainly a function of temperature, crosslinking, and the amount of absorbed penetrant (e.g., swelling in water). The diffusion is pictured as the permeating molecule jumping from one unoccupied volume to another. The jumps are possible due to the thermal motion of the polymer chains. Higher temperatures and less crosslinking increase the thermal motion, while lower temperatures and more crosslinking decrease thermal motion. When the temperature increases, the free volume increases and diffusion in the coatings is faster. When the number of crosslinks increases, the free volume decreases and the diffusion is slower.

10.2.2 Water

The water uptake in coatings depends on several properties of the paint, for example, type of binder and pigmentation. Protective organic coatings have been found to absorb about 2%–5% water [31–33]. The water in the coating may exist in two different chemical states. Either the water molecules are concentrated in clusters, or they are associated with polar groups on the polymer chains, such as hydroxyl groups, amines, or acids. Studies have shown that water does not diffuse into the film uniformly. First, it diffuses into hydrophilic regions and from these into the rest of the polymer [34]. The polymer is not uniform in structure. It can rather be considered as consisting of microgels (high molecular weight and high crosslinking [HMW/HC] density) connected via a low-molecular-weight and low-crosslinked (LMW/LC) polymer fraction. The inhomogeneity is due to a phase separation during film formation [34–36]. The film formation is not a homogeneous process. Formation of microgels starts at a number of different sites in the wet film. As the microgels approach each other from their initiation sites, they are unable to merge into a homogeneous structure. When the microgels meet, the polymerization is terminated, and unreacted or partly reacted resin is left at the periphery. This unreacted or partly reacted resin is the LMW/LC fraction described above. The LMW/LC fraction takes up a large amount of water, has a low resistance to ion transport, and is susceptible to water attack, for example, hydrolysis and dissolution [36]. This inhomogeneity also has implications for the transport of ions through coatings, which will be discussed in Section 10.2.3.

The addition of pigments to the paint can either reduce or increase the water permeation, depending on the paint [28]. The pigments may decrease water permeation by reducing the total permeable volume of the coating or by lengthening the diffusion pathway. An increased accumulation of water at the pigment–binder interface, or an increased amount of voids in the film caused by the pigments, may increase the water permeability. A decrease in the water permeation requires that the pigments are impermeable. Paint producers have used flake-shaped pigments of aluminum, micaceous iron oxide, or glass. The flakes have a tendency to orient themselves parallel to the substrate (Figure 10.2), so the water has to take a longer, more circuitous route around the particles to reach the substrate [20,31]. The orientation parallel to the substrate is

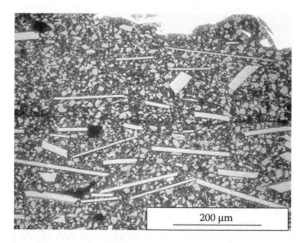

FIGURE 10.2   Glass flake pigments in an epoxy coating, increasing diffusion path through the film.

probably due to flow-out of the wet paint droplets when they hit the substrate. The paint will flow parallel to the substrate, aligning the flakes in the same orientation.

10.2.3 IONS

The transport of ions through organic coatings has received much attention from researchers, due to its direct correlation with coating failure. For a general review about the mathematics and kinetics of the diffusion of ions in polymers, Zaikov et al.'s *Diffusion of Electrolytes in Polymers* is helpful [37]. The important difference between transport of ions and nonionic species (water and oxygen) is that ions are charged. This means that they have very strong interactions with each other, polar groups in the polymer, and polar molecules like water. They are also affected by electric fields. The transport of one ion through a coating is therefore strongly dependent on the transport of other ions, the presence of water, the nature of the polymer, potential gradients, and the solutions outside the coating. All this is dealt with in the theory of irreversible thermodynamics [38].

The mechanism by which ions are transported through organic coatings has been the subject of much research. Several researchers have emphasized the effect of coating inhomogeneity on the transport of ions [36,39–41]. How the inhomogeneity arises was discussed above in connection with the transport of water in coatings. Mayne studied the conductivity of various organic coatings [39] and found that the coating resistance varied with the electrolyte concentration (concentration of ions in the solution) in two different ways. Either the conductivity decreased with increasing concentration (inverse or I-conductivity), or the conductivity increased when the concentration increased (direct or D-conductivity). He found that the coatings he studied were inhomogeneous, that is, containing a mosaic of I- and D-conductivity areas, and related this to inhomogeneity in the crosslinking of the binder. If the more crosslinked areas only take up water, their conductivity will follow the activity of the water in the solution (I-type). If ionic permeation is possible in the less crosslinked areas, the conductivity of these portions will follow that of the external solution (D-type). The coatings studied were less than 100 µm thick. The discovery of the I/D properties was an important finding because it explains why an apparently intact coating can fail, for example, by blistering. Nguyen et al., for example, have used the I/D-theory to explain the development of blisters [36].

Mills and Mayne found that the amount of D-type areas in the film decreases when the film thickness increases or the film is applied in two or more coats [40]. They assumed that the diameter of the D areas is probably small, possibly about 75–250 µm. Thus, when applying the film in two or more coats, the chance for D areas in the two coats to overlap is small and the amount of D areas will decrease. The films they studied were between 40 and 80 µm thick.

Steinsmo studied the conductivity of many different organic coatings of thickness between 100 and 250 µm [42,43]. She found that the resistivity of the coatings varied between 10^9 and 10^{13} Ωcm. For the sample area used, Steinsmo calculated that one pore through the sample with a 10 µm diameter would give a resistivity of about 5.8×10^8 Ωcm. Since this value is much lower than the resistances observed, she concluded that the films studied usually do not have pores. Steinsmo also studied the effect of aluminum pigments and extender pigments on the coating resistance and found that for epoxy- and polyurethane binders, the resistance decreased with increasing aluminum concentration [43]. A study of coatings with constant pigment volume concentration (PVC) but varying aluminum content indicated that the effect was not specific for the aluminum pigments but rather due to a general increase in the PVC. The decrease in resistance for the epoxy was assumed to be due to an increase in ion mobility caused by changes in the polymer network by the pigment particles.

10.2.4 Oxygen

Oxygen is a small nonpolar molecule, and the oxygen molecule therefore has smaller thermodynamic interactions with the polymer chains. The diffusion of oxygen is more ideal then, in the sense that it follows Fick's and Henry's laws to a larger extent than water does. The oxygen diffusion coefficient has been found to be about 10^{-8} cm^2/s in both epoxy and alkyd paint [26,31]. This is about two orders of magnitude higher than the diffusion coefficient of water in the same epoxy [31]. The concentration of oxygen in the alkyd film was found to be about 8×10^{-4} mol/L [26], which is approximately 10% of the concentration in air.

10.2.5 Importance of Barrier Properties

As stated in the introduction to this section, the coating must be a barrier to ions in order to be able to provide durable protection. Water and oxygen, on the other hand, will be absorbed in protective coatings, but will normally cause no harm. In Chapter 12, we will discuss how ions may penetrate coatings and the types of degradation they will cause.

Figure 10.3 shows the resistivity of a polysiloxane coating as a function of film thickness. At about 60 μm, the resistivity suddenly increases by about three orders

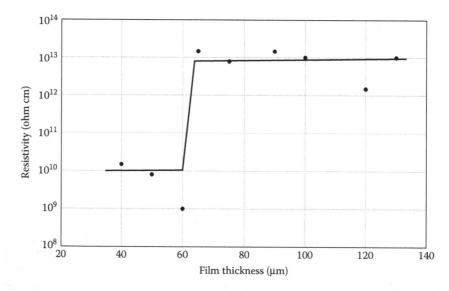

FIGURE 10.3 Resistivity of a polysiloxane single-coat film as function of film thickness.

of magnitude. It is reasonable to interpret this in light of the theories presented above on inhomogeneity in paint films and I/D areas. Below 60 μm, LMW/LC areas may have penetrated the film, resulting in a certain amount of D areas and "low" resistivity in the coating. Above 60 μm, the HMW/HC microgels may have managed to block the D areas. Figure 10.3 is a very good illustration of the importance of sufficient film thickness for protective coatings.

REFERENCES

1. Dickie, R.A. *Prog. Org. Coat.* 25, 3, 1994.
2. Koehler, E.L. Corrosion under organic coatings. In *U.R. Evans International Conference on Localized Corrosion*, Williamsburg, VA, December 1971, pp. 117–133.
3. Troyk, P.R., et al. Humidity testing of silicone polymers for corrosion control of implanted medical electronic prostheses. In *Polymeric Materials for Corrosion Control*, ed. R.A. Dickie and F.L. Floyd. Washington, DC: American Chemical Society, 1986, p. 299.
4. Mittal, K.L. *Adhesion Aspects of Polymeric Coatings*. London: Plenum Press, 1983.
5. Lee, L.H. The chemistry and physics of solid adhesion. In *Fundamentals of Adhesion*, ed. L.H. Lee. London: Plenum Press, 1991, p. 1.
6. Kollek, H., and C. Matz. *Adhesion* 33, 27, 1989.
7. Chruściel, J.J., and E. Leśniak. *Prog. Polym. Sci.* 41, 67, 2015.
8. Bolger, J. Acid base interactions between oxide surfaces and polar organic compounds. In *Adhesion Aspects of Polymeric Coatings*, ed. K.L. Mittal. London: Plenum Press, 1983, p. 3.
9. Mills, G.D. Presented at CORROSION/1986. Houston: NACE International, 1986, paper 313.
10. Hagen, C.H.M., et al. To be published.
11. Hagen, C.H.M., et al. The effect of surface profile on coating adhesion and corrosion resistance. Presented at CORROSION/2016. Houston: NACE International, 2016, paper 7518.
12. Baldan, A. *Int. J. Adhesion Adhesives* 38, 95, 2012.
13. Islam, M.S., et al. *Int. J. Adhesion Adhesives* 51, 32, 2014.
14. Baldan, A. *J. Mater. Sci.* 39, 1, 2004.
15. van der Leeden, M.C., and G. Frens. *Adv. Eng. Mater.* 4, 280, 2002.
16. Piens, M., and H. de Deurwaerder. *Prog. Org. Coat.* 43, 18, 2001.
17. Wenzel, R.N. *Ind. Eng. Chem.* 28, 988, 1936.
18. Watts, J.F., and J.E. Castle. *J. Mater. Sci.* 19, 2259, 1984.
19. Bikerman, J.J. Problems in adhesion measurement. In *Adhesion Measurement of Thin Films, Thick Films, and Bulk Coatings*, ed. K.L. Mittal. Philadelphia: American Society for Testing and Materials, 1978, p. 30.
20. Funke, W. *J. Coat. Technol.* 55, 31, 1983.
21. Funke, W. How organic coating systems protect against corrosion. In *Polymeric Materials for Corrosion Control*, ed. R.A. Dickie and F.L. Floyd. Washington, DC: American Chemical Society, 1986, p. 222.
22. Leidheiser, H. Mechanisms of de-adhesion of organic coatings from metal surfaces. In *Polymeric Materials for Corrosion Control*, ed. R.A. Dickie and F.L. Floyd. Washington, DC: American Chemical Society, 1986, p. 124.
23. Funke, W. In *Surface Coatings—2*, ed. A.D. Wilson, J.W. Nicholson, and H.J. Prosser. London: Elsevier Applied Science, 1988, p. 107.

24. LeBozec, N., et al. Round-robin evaluation of ISO 20340 Annex A test method. Presented at CORROSION/2016. Houston: NACE International, 2016, paper 6991.
25. Meis, N.N.A.H., et al. *Prog. Org. Coat.* 77, 176, 2014.
26. Stratmann, M., et al. *Electrochim. Acta* 39, 1207, 1994.
27. Alexander, M.R., et al. *Surf. Interface Anal.* 29, 468, 2000.
28. Crank, J., and G.S. Park. *Diffusion in Polymers*. London: Academic Press, 1968.
29. Park, G.S. *Transport in Polymer Films*. Dallas: Marcel Decker, 1976.
30. Kumins, C.A., and T.K. Kwei. Free volume and other theories. In *Diffusion in Polymers*, ed. J. Crank and G.S. Park. London: Academic Press, 1968, p. 107.
31. Knudsen, O.Ø., and U. Steinsmo. *J. Corros. Sci. Eng.* 2, 37, 1999.
32. Funke, W. *Ind. Eng. Chem. Prod. Res. Dev.* 24, 343, 1985.
33. Parks, J., and H. Leidheiser. *Ind. Eng. Chem. Prod. Res. Dev.* 25, 1, 1986.
34. Karyakina, M.I., and A.E. Kuzmak. *Prog. Org. Coat.* 18, 1990.
35. González, M.G., et al. Applications of FTIR on epoxy resins—Identification, monitoring the curing process, phase separation and water uptake. In *Infrared Spectroscopy—Materials Science, Engineering and Technology*, ed. T. Theophile. Rijeka, Croatia: InTech, 2012.
36. Nguyen, T., et al. *J. Coat. Technol.* 68, 45, 1996.
37. Zaikov, G.E., et al. *Diffusion of Electrolytes in Polymers*. Utrecht, the Netherlands: VSP, 1988.
38. Førland, K.S., et al. *Irreversible Thermodynamics, Theory and Applications*. New York: John Wiley & Sons, 1988.
39. Mayne, J.E.O. *Br. Corros. J.* 5, 106, 1970.
40. Mills, D.J., and J.E.O. Mayne. The inhomogenous nature of polymer films and its effect on resistance inhibition. In *Corrosion Control by Organic Coatings*, ed. H. Leidheiser. Houston: NACE International, 1981, p. 12.
41. Wu, C., et al. *Prog. Org. Coat.* 25, 379, 1995.
42. Steinsmo, U., and E. Bardal. *J. Electrochem. Soc.* 136, 3588, 1989.
43. Steinsmo, U., et al. *J. Electrochem. Soc.* 136, 3583, 1989.
44. Sørensen, P. A. et al., *Prog. Org. Coat.* 64, 142, 2009.

11 Weathering and Aging of Paint

This chapter presents a brief overview of the major mechanisms that cause aging and subsequent failure of organic coatings. Even the best organic coatings, properly applied to compatible substrates, eventually age when exposed to weather, losing their ability to protect the metal.

In real-life environments, the aging process that leads to coating failure can generally be described as follows:

1. Weakening of the coating by significant amounts of bond breakage within the polymer matrix. Such bond breakage may be caused chemically (e.g., through hydrolysis reactions, oxidation, or free radical reactions) or mechanically (e.g., through freeze–thaw cycling, which leads to alternating tensile and compressive stresses in the coating).
2. Overall barrier properties may be decreased as bonds are broken in the polymeric backbone—in other words, as transportation of water, oxygen, and ions through the coating increases. The polymeric network may be plasticized by absorbed water, which softens it and makes it more vulnerable to mechanical damages. The coating may begin to lose small, water-soluble components, causing further damage.
3. Transportation of ions through the coating.
4. Deterioration of coating–metal adhesion at this interface.
5. Development of internal stress in the coating with subsequent cracking.
6. Development of an aqueous phase at the coating–metal interface and activation of the metal surface for the anodic and cathodic reactions.
7. Corrosion and delamination of the coating.

Many factors can contribute in various degrees to coating degradation, such as

- Ultraviolet (UV) radiation
- Water and moisture uptake
- Elevated temperatures
- Chemical damage (e.g., from pollutants)
- Leaching of components from the film
- Molecular and singlet oxygen
- Ozone
- Abrasion or other mechanical stresses

The major weathering stresses that cause degradation of organic coatings are the first four in the list above: UV radiation, moisture, heat, and chemical damage. And, of course, interactions between these stresses are to be expected; for example, as the polymeric backbone of a coating is slowly being broken down by UV light, the coating's barrier properties can be expected to worsen. Ranby and Rabek [1] have shown that under UV stress, polyurethanes react with oxygen to form hydroperoxides, and that this reaction is accelerated by water. Another example is the temperature–condensation interaction. Elevated temperatures by themselves can damage a polymer; however, they can also create condensation problems, for example, if high daytime temperatures are followed by cool nights. These day–night (diurnal) variations in temperature determine how much condensation occurs, as the morning air warms up faster than the steel.

Various polymers, and therefore, coating types, react differently to changes in one or more of these weathering stresses. Therefore, in order to predict the service life of a coating in a particular application, one must know not only the environment—average time of wetness, amounts of airborne contaminants, UV exposure, and so on—but also how these weathering stresses affect the particular polymer [2].

11.1 UV BREAKDOWN

Sunlight is the worst enemy of paint. It is usually associated with aesthetic changes, such as yellowing, color change or loss, chalking, gloss reduction, and lowered distinctness of image. More important than the aesthetic changes, however, is the chemical breakdown and worsened mechanical properties caused by sunlight. The range of potential damage is enormous [3–7] and includes

- Embrittlement
- Increased hardness
- Increased internal stress
- Generation of polar groups at the surface, leading to increased surface wettability and hydrophilicity
- Changed solubility and crosslink density

In terms of coating performance, this translates into alligatoring, checking, crazing, and cracking; decreased permeation barrier properties; loss of film thickness; and delamination from the substrate or underlying coating layer.

All the damage described above is created by the UV component of sunlight. UV light is a form of energy. When this extra energy is absorbed by a chemical compound, it makes bonds and break bonds. Visible light does not contain the energy required to break the carbon–carbon and carbon–hydrogen bonds most commonly found on the surface of a cured coating. However, just outside of the visible range light in the wavelength range of 285–390 nm contains considerably more energy, commonly enough to break bonds and damage a coating. The 285–390 nm range causes almost all weathering-induced paint failure down at ground level [4]. At the short end of the UV range, we find the most destructive radiation. The damage caused by short-wave radiation is limited, though, to the topmost surface layers of the coating. Longer-wave UV radiation penetrates the film more deeply, but causes less damage [8–10]. This leads to an inhomogeneity in the coating, where the top surface can be more highly crosslinked than the bulk of the coating layer [4]. As the top surface of the film eventually breaks up, chalking and other degradation phenomena become apparent. (The light located below 285 nm, with even higher energy, can easily break carbon–carbon and carbon–hydrogen bonds and has enough energy left over for considerably more mischief as well. However, the Earth's atmosphere absorbs most of this particular wavelength band of radiation, and therefore, it is a concern only for aircraft coatings, which receive less protection from the ozone layer.)

The interactions of coatings with UV radiation may be broadly classed as follows:

- Light is reflected from the film.
- Light is transmitted through the film.
- Light is absorbed by a pigment or by the polymer.

In general, reflectance and transmittance pose no threats to the life span of the coating. Absorption is the problem. When energy from the sun is absorbed, it leads to chemical destruction (see Section 11.1.3).

11.1.1 Reflectance

Light is reflected from the film by the use of leafy or plate-like metal pigments located at the top of the coating. These are surface treated so that the binder solution has difficulty wetting them. When the film is applied, the plate-like pigments float to the top of the wet film and remain there throughout the curing process. The dried film has a very thin layer of binder on top of a layer of pigment that is impermeable to light. The binder on top of the pigment layer may be broken down by UV radiation and disappear, but as long as the leafy pigments can be held in place, the bulk of the binder behind the leafy pigments is shielded from sunlight.

11.1.2 Transmittance

Transmitted light, which passes through the film without being absorbed, does not affect the structure of the film. Of course, if a coating layer underneath is sensitive to UV radiation, problems can occur. Epoxy coatings, which are the most important class of anticorrosion primer and barrier, are highly sensitive to UV radiation. Epoxies are therefore generally covered by a topcoat, whose main function is to not transmit the UV radiation and protect the epoxy.

11.1.3 Absorption

Light can be absorbed by a pigment, the binder, or an additive. Light absorbed by the pigment is dissipated as heat, which is a less destructive form of energy than UV light [4]. The real damage comes from the UV radiation absorbed by the nonpigment components of the coating—that is, the polymeric binder.

UV energy absorbed by the binder can wreak havoc in wild ways. The extra energy can go into additional crosslinking of the polymer, or it can start breaking the existing bonds.

Because the polymer chains in the cured film are well anchored and already crosslinked, further crosslinking results in additional tightening of the polymer chains [7]. This increases the internal stress of the cured film, which in turn leads to hardening, decreased flexibility, and embrittlement. If the internal stresses overcome the cohesive strength of the film, then the unfortunate end is cracking; if failure takes the form of lost adhesion at the coating–metal interface, then delamination is seen. Both, of course, can happen simultaneously.

Instead of causing additional crosslinking, the UV energy could break bonds in the polymer or another component of the coating. Free radicals are thus initiated. These free radicals react with either

- Oxygen to produce peroxides, which are unstable and can react with polymer chains
- Other polymer chains or coating components to propagate more free radicals

Reaction of the polymer chain with peroxides or free radicals leads to chain breaking and fragmentation. *Scissoring*, a term used to describe this reaction, is an apt description. The effect is exactly as if a pair of scissors was let loose inside the coating, cutting up the polymer backbone. The destruction is enormous. When scissoring cuts off small molecules, they can be volatilized and make their way out of the coating. The void volume necessarily increases as small parts of the binder disappear (and, of course, ultimately the film thickness decreases). The internal stress on the remaining anchored polymer chains increases, leading to worsened mechanical properties. After enough scissoring, the crosslink density has been significantly altered for the worse, loss of film thickness occurs, and a decrease in permeation barrier properties is seen. The destruction stops only when two free radicals combine with each other, a process known as *termination* [4,11].

Table 11.1 summarizes the effects on the coating when absorbed UV energy goes into additional crosslinking, scissoring, or generating polar groups at the coating surface.

Ideally, a selection of binders that absorb little or no UV radiation should minimize the potential damage from this source. In reality, however, even paints based on these

TABLE 11.1
Effects of Absorbed UV Energy

Absorbed UV Energy Goes Into	Which Causes	And Ultimately
Additional crosslinking	Increased internal stress, leading to hardening, decreased flexibility, and eventually embrittlement	Cracking, delamination, or both
Scissoring	Increased internal stress, increased void volume, and worsened crosslink density	Loss of film thickness and decrease of permeation barrier properties
Generation of polar groups at the surface	Increased surface wettability and hydrophilicity	Decrease of permeation barrier properties

binders can prove vulnerable because other components—both those intentionally added and those that were not—often compromise the coating as a whole. Components that can be said to have been added intentionally are, of course, pigments and various types of additives: antiskinning, antibacterial, emulsifying, colloid-stabilizing, flash-rust-preventing, flow-controlling, thickening, viscosity-controlling additives, and so forth. Examples of unintentional components are catalysts or monomer residues left over from the polymer processing; these may include groups that are highly reactive in the presence of UV radiation, such as ketones and peroxides. Interestingly, impurities can sometimes show a beneficial effect. When studying waterborne acrylics, Allen and colleagues [12] have found that low levels of certain comonomers reduced the rate of hydroperoxidation. The researchers speculate that the styrene comonomer reduced the unzipping reaction that the UV otherwise would cause.

11.2 MOISTURE

Moisture (water or water vapor) can come from several sources, including water vapor in the surrounding air, rain, and condensation as temperatures drop at night. Paint films constantly absorb and desorb water to maintain equilibrium with the amount of moisture in the environment. Water is practically always present in the coating. In a study of epoxy, chlorinated rubber, alkyd, and linseed oil paints, Lindqvist [13] found that even in stagnant air at 25°C and 20% relative humidity (RH), the smallest equilibrium amount of water measured was 0.04 wt%.

Water or water vapor is taken up by the coating as a whole through pores and microcracks; the binder itself also absorbs moisture. Water uptake is not at all homogeneous; it enters the film in several different ways and can accumulate in various places [13,14]. Within the polymer phase, water molecules can be randomly distributed or aggregate into clusters, can create a watery interstice between binder and pigment particle, can exist in pores and voids in the paint film, and can accumulate at the metal–coating interface. Once corrosion has begun, water can exist in blisters or in corrosion products at the coating–metal interface.

Water molecules can exist within the polymer phase because polymers generally contain polar groups that chemisorb water molecules. The chemisorbed molecules can be viewed as bound to the polymer because the energy for chemisorption (10–100 kcal/mol) is similar to that required for chemical bonding. The locked, chemisorbed molecule can be the center for a water cluster to form within the polymer phase [13].

When water clusters form in voids or defects in the film, they can behave as fillers, stiffening the film and causing a higher modulus than when the film is dry. Funke and colleagues [14] concluded that moisture in the film can have seemingly contradictory effects on the coating's mechanical properties because several different—and sometimes opposite—phenomena are simultaneously occurring.

Two of the most important parameters of water permeation are solubility and diffusion. Solubility is the maximum amount of water that can be present in the coating in the dissolved state. Diffusion is how mobile the water molecules are in the coating [15]. The permeability coefficient, P, is the product of the diffusion coefficient, D, and the solubility, S [16]:

$$P = D \times S$$

In accelerated testing, the difference in absorption and desorption rates of water for various coatings is also important (see Chapter 14).

The uptake of water affects the coating in several ways [17]:

- Chemical breakdown
- Weathering interactions
- Hygroscopic stress
- Blistering and adhesion loss

11.2.1   CHEMICAL BREAKDOWN AND WEATHERING INTERACTIONS

Water is an excellent solvent for atmospheric contaminants, such as salts, sulfites, and sulfates. Airborne contaminants would probably never harm coated metals, if not for the fact that they so easily become Cl^- or SO_4^{2-} ions in water. The water and ions, of course, fuel corrosion beneath the coating. Water can also be a solvent for some of the additives in the paint, causing them to dissolve or leach out of the cured film. And finally, it can act as a plasticizer in the polymeric network, softening it and making it more vulnerable to mechanical damages. But in general, protective organic coatings must be chemically stable in water, since water or humidity is an essential component in order to make an environment corrosive. Additional stresses are needed in order for water to be able to attack a protective organic coating.

As previously noted, the major weathering stresses interact with each other. Perera and colleagues have shown that temperature effects are inseparable from the effects of water [18,19]. The same is even more true for chemical effects.

The effects of UV degradation can be worsened by the presence of moisture in the film [1]. As a binder breaks down due to UV radiation, water-soluble binder fragments can be created. These dissolve when the film takes up water, are removed from the film upon drying, and add to the decrease in film density or thickness.

11.2.2   HYGROSCOPIC STRESS

This section focuses on the changes in the coating's internal stresses—both tensile and compressive—caused by wetting and drying the coating. Volume changes in the film cause internal stress, either compressive or tensile, because adhesion to the substrate prevents lateral movement. The film is three-dimensional, but can only change in thickness. As a coating takes up water, it swells, causing compressive stresses in the film. As the coating dries, it contracts, causing tensile stress. These compression and tension forces have adverse effects on the film's cohesive integrity and on its adhesion to the substrate. Of the two types of stresses, the tensile stresses formed as the coating dries have the greater effect [9,11,20–22].

Coating stress is a dynamic phenomenon; it changes drastically during water uptake and desorption. Sato and Inoue [23] have reported that the initial tensile stresses (left over from shrinkage during film formation) of the dry film decrease to zero as moisture is absorbed. Once the initial tensile stresses have been negated by water uptake, further uptake leads to buildup of compressive stresses. If the film is dried, tensile (shrinkage) stresses redevelop, but to a lower degree than originally seen. Some degree of permanent creep was seen in Sato and Inoue's study; it was attributed to the breaking and reforming of valency associations in the epoxy polymer. The same trend of initial tensile stress reduction, followed by compressive stress buildup, was seen by Perera and Vanden Eynde [24] with a polyurethane and a thermoplastic latex coating.

Hygroscopic stresses are interrelated with ambient temperature [11,19]. They also depend heavily on the glass transition temperature (T_g) of the coating [25]. In immersion studies, Perera and Vanden Eynde examined the stress of an epoxy coating whose T_g was near—even below—the ambient temperature [26]. The films in question initially had tensile stress from the film formation. Upon immersion, this stress gradually disappeared. As in the previously cited studies, compressive stresses built up. The difference was that these stresses then dissipated over several days even though immersion continued. Hare also noted dissipation of compressive stresses as the difference between $T_{ambient}$ and T_g is reduced; he attributes this to a reduced modulus and a flexibilizing of the film [11]. Because of the low T_g of the film, stress relaxation occurred and the compressive stresses due to water uptake disappeared.

Hygroscopic stresses have a very real effect on coating performance. If a coating forms high levels of internal stress during cure—not uncommon in thick, highly crosslinked coatings—then applying other stresses during water uptake or desorption can lead to cracking or delamination. Hare has reported another problem: cases where the film expansion during water uptake created a strain beyond the film's yield point. Deformation here is irreversible; during drying, permanent wrinkles are left in the dried paint [17]. Perera et al. have pointed out that hygroscopic stress can be critical to designing accelerated tests for coatings. For example, a highly crosslinked coating can undergo more damage during the few hours it dries after the salt spray test has ended than it did during the entire time (hundreds of hours) of the test itself [27].

An additional effect of water that affects internal stress is that components of the film can be washed out [28]. Most coatings contain small quantities of molecules that are lost during exposure in wet environments. This will cause additional volume change in the film and increase the level of internal stress.

11.2.3 BLISTERING

Blistering is not, strictly speaking, brought about by aging of the coating. It would be more correct to say that blistering is a sign of failure of the coating–substrate system. Blistering occurs when moisture penetrates through the film and accumulates at the coating–metal interface in sufficient numbers to force the film up from the metal substrate. The two types of blistering in anticorrosion paints—alkaline and neutral—are caused by different mechanisms.

Blistering is also closely correlated with ions penetrating the film, as discussed in Section 10.2.3.

11.2.3.1 Alkaline Blistering

Alkaline blistering occurs when cations, such as sodium (Na^+), migrate to cathodic areas at the coating–metal interface via coating defects, such as pores, scratches, or any area with inferior barrier properties, for example, low film thickness. At the cathodic areas, the cations combine with the hydroxyl anions produced by the cathodic reaction to form sodium hydroxide (NaOH). The result is a strongly alkaline aqueous solution at the cathodic area. As osmotic forces drive water through the coating to the alkaline solution, the coating is deformed upward—a blister begins. At the coating–metal solution interface, the coating experiences peel forces, as shown in Figure 11.1. It is well established that the force needed to separate two adhering bodies is much lower in peel geometry than in the tensile geometry normally used in adhesion testing of coatings. At the edge of the blister, the coating may be adhering as tightly as ever to the steel. However, because the coating is forced upward at the blister, the coating at the edge is now undergoing peeling and the force needed to detach the coating in this geometry is lower than the forces measured in adhesion tests. Cathodic disbonding also breaks the bonds between the coating and the substrate (Section 12.1). This facilitates growth of the blisters until the film cracks and the osmotic pressure is released.

Leidheiser and colleagues [29] have shown that cations may diffuse laterally via the coating–metal interface. Their elegantly simple experiment demonstrating this is shown in Figure 11.2. Adhesion is significantly less under wet conditions (see Section 10.1.4), making ion migration along the interface easier. Lateral diffusion along the coating–metal interface has since been confirmed by other experimental techniques [30,31].

11.2.3.2 Neutral Blistering

Neutral blisters contain solution that is weakly acid to neutral. No alkali cations are involved. The first step is undoubtedly reduction of adhesion due to water clustering at the coating–metal interface. Funke [32] postulates that differential aeration is responsible for neutral blistering. The steel under the water does not have as ready access to oxygen as the adjacent steel, and polarization arises. The oxygen-poor center of the blister becomes anodic, and the periphery is cathodic. Funke's mechanism

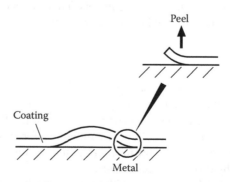

FIGURE 11.1   Peel forces at the edge of a blister.

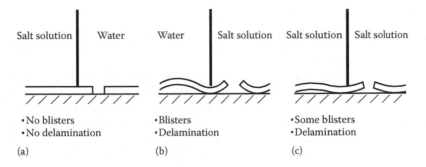

FIGURE 11.2 Experiment of Leidheiser et al. establishing the route of cation diffusion. (From Leidheiser, H., et al., *Prog. Org. Coat.*, 11, 19, 1983.)

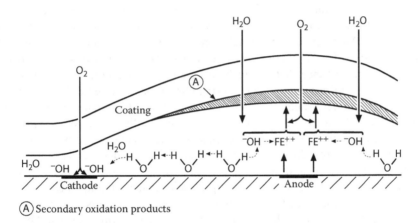

FIGURE 11.3 Mechanism of neutral blistering. (From Funke, W., *Ind. Eng. Chem. Prod. Res. Dev.*, 24, 343, 1985.)

of neutral blistering is shown in Figure 11.3. Painting on salt-contaminated surfaces will inevitably also lead to osmotic blistering.

11.3 TEMPERATURE

In general, ambient temperature changes can alter

- Balance of stresses in the coating–substrate system
- Mechanical properties of the viscoelastic coating
- Barrier properties of the coating

The balance of stresses is affected in various ways. At slightly elevated temperatures, crosslinking of the polymer will continue, making the film stiffer and harder. If the coating is exposed at too high temperature, bonds that are needed begin to break, and the polymer is weakened. Differences in coefficients of thermal expansion also cause thermal stress; epoxies or alkyds, for example, typically have a coefficient of thermal expansion that is twice that of aluminum or zinc and four times that of steel [33]. Whereas too high of a temperature typically causes unwanted chemical changes in the film, very low temperature will create internal tensile stress due to thermal expansion differences [34].

Another factor that must be considered at elevated temperatures is the glass transition temperature (T_g) of the polymer used in the binder. This is the temperature above which the polymer exists in a rubbery state and below which it is in the glassy state. Using coatings near the T_g range is problematic, because the binder's most important properties change in the transition from glassy to rubbery. For example, above the T_g, polymer chain segments undergo Brownian motion. Segments with appropriate functional groups for bonding are increasingly brought into contact with the metal surface. An increase in the number of bond sites can dramatically improve adhesion; wet adhesion in particular can be much better above the T_g than below it.

Increased Brownian motion is also associated with negative effects, such as increased diffusion. Above the T_g, the Brownian motion gives rise to the continuous appearance and disappearance of small pores, 1–5 nm or smaller, within the binder matrix. The size of these small pores compares to the "jump distance" of diffusing molecules—the distance that has to be covered by a molecule moving from one potential energy minimum to a neighboring one in the activated diffusion process. The permeation rate through these small pores is linked to temperature to the same degree that the chain mobility is. That is, the chain mobility of elastomeric polymers shows a high degree of temperature dependence and thus favors activated diffusion at higher temperatures. As the crosslink density of the binder increases, segmental mobility decreases, even at elevated temperatures. Diffusion still occurs through large pore systems whose geometry is largely independent of temperature. The temperature dependence of diffusion in highly crosslinked binders is a result of the temperature dependence of the viscous flow of the permeating species. Above T_g, coatings have so poor barrier properties that they cannot be considered protective.

Miszczyk and Darowicki have found that the increased water uptake at elevated temperatures can be to some extent irreversible; the absorbed water was not fully desorbed during subsequent temperature decreases. They speculate that the excess water may be permanently located in microcracks, microvoids, and local delamination sites [33].

11.4　CHEMICAL DEGRADATION

All breakdowns in polymers could, of course, be regarded as chemical degradation of some sort. What is meant here by the term *chemical degradation* is breakdown in the paint film that is induced by exposure to chemical contaminants in the atmosphere.

Atmospheric contaminants play a more minor role in polymer breakdown than do UV exposure, moisture, and (to a lesser degree) temperature. However, they can contribute to coating degradation, especially when they make the coating more vulnerable to degradation by UV light, water, or heat.

Chemical species, such as road salts and atmospheric contaminants in the wind or rain, are routinely deposited on paint surfaces. There, they combine with condensation to form aggressive, usually saline or acidic solutions. Most polymers used in modern coatings have good resistance to acids and salts; however, modern coatings also contain a large number of additives, which can prove vulnerable to chemical attack. For example, many coatings contain light stabilizers based on hindered amines to aid UV resistance. It is well known that the performance of these stabilizers is diminished by acids and pesticides [35]. When this occurs, chemical exposure makes the coating vulnerable to UV breakdown.

Chemical degradation inevitably boils down to selecting a coating that is able to withstand the specific exposure environment. The performance of a coating will differ from place to place. We tend to think of field testing of coatings as providing the ultimate answer to coating lifetime expectancy, but exposure sites differ with respect to weather conditions, UV radiation, and atmospheric pollution. Sampers [35] reports that in a study of polyolefin samples exposed both in Florida and on the Mediterranean coast of France, a dramatic difference was seen in polymer lifetime. Samples exposed on the Mediterranean had only half the life of those in Florida. The two stations had broadly similar weathering parameters; the differences should have led to *longer* lifetimes in France. Sampers concluded that constituents in the rain or wind had chemically interacted with the hindered amine light stabilizers in the polymers exposed in France, causing these samples to be especially vulnerable to UV degradation.

In a study of gloss retention of coatings exposed for 2½ years at weathering sites in Kuwait, Carew and colleagues [36] reported a probable link between industrial pollution and coating damage, although in this case the damage seems to have been caused by dust from a cement factory. The sites in this study are described in Table 11.2. Because all the sites are located in the Shuaiba area of Kuwait's industrial belt, they should be very similar in temperature and humidity. The difference between sites is the distance from the Arabian Gulf and the amount and type of atmospheric pollution. Carew and colleagues found that coatings consistently showed the

TABLE 11.2
Description of Sites in Kuwait Study

Site	Distance from Sea (km)	Pollution	Notes
A	0.2	Heavy	Downwind from refinery and salt and chlorine plant
B	0.55	Heavy	Next to refinery and desalination and electricity production plant
C	1.5	Heavy	Upwind from refinery, next to cement clinker factory
D	3	Mild	Rural area

Source: Data from Carew, J.A., et al., Weathering performance of industrial atmospheric coatings systems in the Arabian Gulf, presented at CORROSION/1994, NACE International, Houston, 1994, paper 445.

worst performance at site C, although this site is farther from the sea than sites A and B and, being upwind, does not suffer from the refinery. However, they also noted heavy amounts of dust on the samples at this site, almost certainly from the cement factory next door. Analysis of the dust showed it to be similar to the composition of clinker cement. Cement, of course, is extremely alkaline, to which few polymers are resistant. At the high temperatures at these sites—up to 49°C—and with the very high amounts of water vapor available, soluble alkaline species in the dust deposits can form a destructive, highly alkaline solution that can break down cured binder. The extent to which the various coatings managed to retain gloss at this site is almost certainly a reflection of the polymer's ability to resist saponification.

In a study of coated panels exposed throughout two pulp and paper mills in Sweden, Rendahl and colleagues [37] found that the amounts of airborne H_2S and SO_2 at the various locations did not have a significant impact on coating performance. The effect of airborne chlorine in this study is not clear; the authors note that only total chlorine was measured, and the amounts of active corrosion-initiating species at each location are unknown.

Özcan and colleagues [38] examined the effects of very high SO_2 concentrations on polyester coatings. Using 0.286 atm SO_2 (to simulate conditions in flue gases) and RH ranging from 60% to 100%, they found that corrosion occurred only in the presence of water. At 60% RH, no significant corrosion damage occurred, despite the very high concentration of SO_2 in the atmosphere.

Another study, performed in Spain, indicates that humidity played a more important role than levels of atmospheric contaminants in predicting corrosion of painted steel [39]. However, without quantitative data of pollutant levels for Madrid and Hospitalet, it is impossible to rule out a combination of humidity and airborne pollutants as the major factor in determining coating performance. In this study, 60 μm chlorinated rubber was applied to clean steel. Painted samples and coupons of bare steel and zinc were exposed in dry rural, dry urban, humid industrial, and humid coastal areas. The results after two years are given in Table 11.3.

TABLE 11.3
Performance of Bare Steel and Coated Panels

Location	Type of Atmosphere	Humid/Dry	Corrosion of Bare Steel (μm/year)	Degree of Oxidation of Painted Surface (%) after Two Years
El Pardo	Rural	Dry	14.7	0
Madrid	Urban	Dry	27.9	0
Hospitalet	Industrial	Humid	52.7	0.3
Vigo	Coastal	Humid	62.6	16

Source: Modified from Morcillo, M., and Feliu, S., in *Proceedings Corrosio i Medi Ambient*, Universitat de Barcelona, Barcelona, 1986, p. 312.

REFERENCES

1. Ranby, B., and J.F. Rabek. *Photodegradation, Photo-Oxidation and Photostabilization of Polymers: Principles and Application.* New York: Wiley Interscience, 1975.
2. Forsgren, A., and C. Appelgren. Performance of organic coatings at various field stations after 5 years' exposure. Report 2001:5D. Stockholm: Swedish Corrosion Institute, 2001.
3. Berg, C.J., et al. *J. Paint Technol.* 39, 436, 1967.
4. Hare, C.H. *J. Prot. Coat. Linings* 17, 73, 2000.
5. Krejcar, E., and O. Kolar. *Prog. Org. Coat.* 3, 249, 1973.
6. Nichols, M.E., and C.A. Darr. *J. Coat. Technol.* 70, 141, 1998.
7. Oosterbroek, M., et al. *J. Coat. Technol.* 63, 55, 1991.
8. Fitzgerald, E.B. *ASTM Bull.* 207 TP-137, 650, 650.
9. Marshall, N.J. *Off. Dig.* 29, 792, 1957.
10. Miller, C.D. *J. Am. Oil Chem. Soc.* 36, 596, 1959.
11. Hare, C.H. *J. Prot. Coat. Linings* 13, 65, 1996.
12. Allen, N.S., et al. *Prog. Org. Coat.* 32, 9, 1997.
13. Lindqvist, S. *CORROSION* 41, 69, 1985.
14. Funke, W., et al. *J. Coat. Technol.* 68, 210, 1996.
15. Hulden, M., and C.M. Hansen. *Prog. Org. Coat.* 13, 171, 1985.
16. Ferlauto, E.C., et al. *J. Coat. Technol.* 66, 85, 1994.
17. Hare, C.H. *J. Prot. Coat. Linings* 13, 59, 1996.
18. Perera, D.Y. *Prog. Org. Coat.* 44, 55, 2002.
19. Perera, D.Y., and D. Vanden Eynde. *J. Coat. Technol.* 59, 55, 1987.
20. Prosser, J.L. *Mod. Paint Coat* 67, 47, 1977.
21. Axelsen, S.B., et al. *CORROSION* 66, 065005, 2010.
22. Korobov, Y., and D.P. Moore. Performance testing methods for offshore coatings: Cyclic, EIS and stress. Presented at CORROSION/2004. Houston: NACE International, 2004, paper 04005.
23. Sato, K., and M. Inoue. *Shikizai Kyosaish* 32, 394, 1959. Summarized in Hare, C.H., *J. Prot. Coat. Linings* 13, 59, 1996.
24. Perera, D.Y., and D. Vanden Eynde. Use of internal stress measurements for characterization of organic coatings. Presented at 16th FATIPEC Congress. Paris: Fédération d'Associations de Techniciens des Industries des Peintures, Vernis, Emaux et Encres d'Imprimerie de l'Europe Continentale (FATIPEC), paper 1982.
25. Perera, D.Y. *Prog. Org. Coat.* 28, 21, 1996.
26. Perera, D.Y., and D. Vanden Eynde. Presented at 20th FATIPEC Congress, Stress in Organic Coatings under Wet Conditions. Paris: Fédération d'Associations de Techniciens des Industries des Peintures, Vernis, Emaux et Encres d'Imprimerie de l'Europe Continentale (20th FATIPEC), paper 1990.
27. Perera, D.Y., et al. *Polym. Mater. Sci. Eng.* 73, 187, 1995.
28. Knudsen, O.Ø., et al. Development of internal stress in organic coatings during curing and exposure. Presented at CORROSION/2006. Houston: NACE International, 2006, paper 06028.
29. Leidheiser, H., et al. *Prog. Org. Coat.* 11, 19, 1983.
30. Stratmann, M. *CORROSION* 61, 1115, 2005.
31. Wapner, K., et al. *Electrochim. Acta* 51, 3303, 2006.
32. Funke, W. *Ind. Eng. Chem. Prod. Res. Dev.* 24, 343, 1985.
33. Miszczyk, A., and K. Darowicki. *Prog. Org. Coat.* 46, 49, 2003.
34. Bjørgum, A., et al. Protective coatings in arctic environments. Eurocorr paper 1506. Frankfurt am Main: Dechema, 2012.

35. Sampers, J. *Polym. Degrad. Stabil.* 76, 455, 2002.
36. Carew, J., et al. Weathering performance of industrial atmospheric coating systems in the Arabian Gulf. Presented at CORROSION/1994. Houston: NACE International, 1994, paper 445.
37. Rendahl, B., et al. Field testing of anticorrosion paints at sulphate and sulphite mills. In *9th International Symposium on Corrosion in the Pulp and Paper Industry.* Quebec: PAPRICAN, 1998.
38. Özcan, M., et al. *Prog. Org. Coat.* 44, 279, 2002.
39. Morcillo, M., and S. Feliu. Quantitative data on the effect of atmospheric contamination in coatings performance. In *Proceedings Corrosio i Medi Ambient.* Barcelona: Universitat de Barcelona, 1986, p. 312.

12 Degradation of Paint by Corrosion

Corrosion reactions under the paint film are the ultimate failure mode of a protective coating, and the processes that cause the most coating repair operations on coated metal constructions. In this chapter, degradation by corrosion is divided into three mechanisms: cathodic disbonding (CD), corrosion creep, and filiform corrosion. CD is the degradation mechanism by which coatings fail on submerged and buried steel. Corrosion creep is the degradation mechanism for protective coatings on atmospherically exposed steel, and therefore the most important one in terms of area, since this constitutes the largest area of protective coatings. CD may play a role in this type of degradation too, as we will see. Filiform corrosion is the primary electrochemical failure mode for coatings on aluminum, but it can also occur on steel and other metals.

12.1 CATHODIC DISBONDING

This is probably the coating degradation mechanism that has received the most attention from researchers, but still there are details about CD that are not fully understood. Here, an overview of CD will first be presented, and then we will go into the details of the mechanism.

CD is a process causing the organic coating to lose adhesion to the substrate. The adhesion loss is caused by cathodic oxygen reduction:

$$O_2 + 2H_2O + 4e^- = 4OH^- \qquad (12.1)$$

The hydroxide or an intermediate reaction species in Reaction 12.1 breaks the bonds between the coating and the substrate. As a result, the coating can easily be peeled off the substrate and an alkaline electrolyte is found in the narrow crevice under the coating [1]. CD is the most important degradation mechanism for organic coatings on submerged steel, but it can also play an important role in corrosion creep, as will be shown in Section 12.2.

Figure 12.1 shows a schematic illustration of the disbonding process on a painted steel surface exposed in seawater, polarized by a zinc anode. Structures in seawater protected by zinc or aluminum anodes are cathodically polarized to potentials between 800 and −1150 mV Ag/AgCl, depending on the anode material and the distance to the anode. At these potentials, the cathodic reaction can be both hydrogen evolution and reduction of oxygen. Under the coating, the potential will be higher; however, due to the resistance in the electrolyte in the narrow crevice between the coating and the steel, an oxygen reduction is the only cathodic reaction. The process

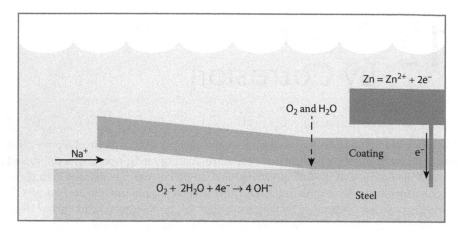

$$Zn = Zn^{2+} + 2e^-$$

$$O_2 \text{ and } H_2O$$

$$Na^+$$

Coating e^-

$$O_2 + 2H_2O + 4e^- \rightarrow 4\,OH^-$$

Steel

FIGURE 12.1   Schematic illustration of CD on a subsea steel surface.

starts at coating edges, mechanical damages, blisters, and so forth, where the steel is exposed to the electrolyte. The cathodic reaction takes place on the exposed steel, and the anodic reaction takes place on the sacrificial anode. Electrons are transported in the steel and the connectors between the anode and the cathode, while ions are transported in the electrolyte to maintain charge neutrality. In seawater, sodium ions are transported under the coating and sodium hydroxide is formed. The coating starts to lose adhesion around the damage, and the disbonded area grows linearly with time [1–4], if all other parameters (potential, temperature, electrolyte, and oxygen concentration) are constant. The pH of the electrolyte under the disbonded coating can be as high as 14 [5,6]. The crevice under the disbonded coating may be only a few micrometers high, so CD is normally not visible on the surface of the coating. Without external polarization, the process is basically the same, but the anodic reaction will be dissolution of the substrate at the coating damage, instead of at the sacrificial anode.

CD has been subject to research for several decades, and several research groups have contributed to our understanding of the phenomenon. Not all aspects of the disbonding process are fully understood, and the deadhesion process has been subject to debate. This is discussed in Section 12.1.2. There has also been disagreement about the transport routes of the reactants to the disbonding front (oxygen, water, and cations), but there now seems to be a general agreement that oxygen and water are transported through the coating, and that cations are transported in the crevice under the disbonded coating.

12.1.1 PARAMETERS AFFECTING CATHODIC DISBONDING

A number of parameters have been shown to affect the rate of CD; a brief overview is given here. Based on the investigations of these parameters, attempts have been made to formulate a unifying degradation mechanism. However, as we will see, there is still some controversy about various aspects about the mechanism, the adhesion loss reaction, and the rate-limiting step in particular. This is discussed in Sections 12.1.2 and 12.1.4.

The disbonding rate is proportional to the potential of the coated steel. CD experiments performed at potentials between −450 and −1450 mV versus Ag/AgCl have shown that the lower the potential is, the higher the disbonding rate will be [2,7–9]. Jin [2] and Steinsmo [8] postulated that the disbonded area was linearly proportional to the potential, while Sørensen found that the disbonded distance was proportional to the potential [9]. There was some scatter in the results presented by Steinsmo, so these results may also be indicative of a linear relationship between distance and potential, as Sørensen found.

The type of electrolyte and its concentration also affect the disbonding rate. Leidheiser et al. found that the disbonding rate increases with cation type in the following order: $CaCl_2$, LiCl, NaCl, KCl, and CsCl, which agrees with the mobility of the cations in water [1,10]. In solutions of divalent cations, CD is very slow or does not occur at all [5,11,12]; this is probably due to the low solubility of the various hydroxides, resulting in a lower pH compared with the alkali metals. CD is assumed to be independent of the type of anion in the solution, since disbonding experiments performed in NaCl, NaBr, and NaF solutions gave the same result [11]. The rate of disbonding has been found to increase with potassium chloride concentration up to about 0.5 M, and then decrease slightly from 0.5 to 2 M concentration. The increased disbonding at low concentrations was attributed to increased conductivity, while the decreased disbonding above 0.5 M was attributed to decreased oxygen solubility.

The oxygen concentration in the electrolyte where the coated steel is exposed affects the disbonding rate. Higher oxygen concentration results in higher disbonding rates, while oxygen-free conditions give low disbonding rates [9,10,13]. This is the case both when the painted steel is cathodically polarized [10,14] and when there is no external polarization [4,9].

Increasing the coating film thickness decreases the CD rate. Two investigations have concluded that the disbonding rate decreased linearly with dry-film thickness [1,2]. Both studies were performed on films with less than 100 μm thickness; a third study, using thicker epoxy coatings, has also concluded that increasing film thickness will decrease disbonding [10].

Increasing surface roughness has been shown to decrease CD [2,15]. In these studies, the surface was characterized by the Ra parameter, that is, average deviation from the centerline in the roughness profile. Ra varied between 0.1 and 3.8 μm, but the effect was not linear in the entire roughness range. The effect was large up to $Ra = 2$. In a later study, CD was shown to be linearly dependent on the tortuosity of the steel surface [16]. Tortuosity is another way of quantifying roughness and is defined as the actual interfacial length, following the surface roughness between two points on a surface, divided by the length of a straight line between the same two points. Tortuosity therefore describes the surface profile better than simply measuring peak height, as the Ra parameter does.

12.1.2 ADHESION LOSS MECHANISM

Various mechanisms for CD have been suggested; these can be classified in three groups, associated with different loci of failure [17]:

1. Dissolution of the iron oxide layer on the substrate
2. Chemical degradation of the coating
3. Interfacial failure caused by the high pH

12.1.2.1 Dissolution of the Iron Oxide Layer on the Substrate

Steel surfaces are covered with oxides, even newly blast-cleaned steel. The composition and stability of these oxides depend on the pH and potential. Loss of adhesion by dissolution of the oxide during cathodic polarization has therefore been suggested [1,6]. Visual changes on the steel surface after disbonding are often observed and can be interpreted as alkaline attack on the oxide film [1]. However, the fact that CD also occurs on stainless steel, copper, brass, zinc, and other metals with stable oxides speaks against this theory [18].

12.1.2.2 Chemical Degradation of the Coating

The oxygen reduction reaction is a multistep reaction involving several unstable intermediates, including free radicals and hydrogen peroxide [19]. It is therefore reasonable to suggest that reactive intermediates may attack the coating binder, and that this may cause the observed disbonding of the coating. Analysis of the coating and steel surface after CD by x-ray photoelectron spectroscopy (XPS) has shown chemical changes in the binder and remnants of the organic coating on the steel surface, which is consistent with disbonding caused by chemical degradation of the binder [20].

Adding different free radical scavengers to an epoxy model coating has been shown to reduce the rate of CD [18]. However, the effect was only found for certain scavengers. The effect cannot be taken as conclusive evidence for the free radical mechanism either, since the scavenger also will prevent formation of hydroxide, the culprit for the other two suggested failure sites. We should therefore expect the scavenger to have the same effect on oxide dissolution and interfacial failure.

12.1.2.3 Interfacial Failure

XPS has been used to analyze the metal surface after CD [21]. In the first 1–2 mm from the scribe, under the disbonded coating, findings were consistent with disbonding caused by metal oxide dissolution, but after that, the locus of failure changed from the metal–oxide junction to the oxide–polymer junction. Thus, the failure was mainly adhesive. Aluminum pigments have been shown to reduce CD significantly when applied in the first coat, although not in subsequent coats [22]. Hence, the aluminum must affect the chemistry that causes the disbonding, not simply by a barrier improvement effect. The aluminum pigments were shown to corrode in the process, indicating that they function as a buffer, removing hydroxide from the steel–coating interface. Aluminum is amphoteric and will corrode at high pH, consuming the hydroxide. This reaction cannot affect the free radical formation, and strongly indicates that hydroxide itself is the substance that causes the disbonding.

12.1.3 Transport of Reactants

In theory, the transport of cations to the disbonding front may go through the coating or in the aqueous film under the disbonded coating. There has been some debate about this, but based on experimental results, we can now firmly conclude that cations are transported under the disbonded coating.

Leidheiser et al. studied the disbonding of a coating in a two-compartment cell under free corrosion potential [1]. The coating was in contact with an electrolyte in one cell and distilled water in the other. When the defect in the coating was in contact with distilled water, no disbonding occurred. When the defect was in contact with the electrolyte, disbonding occurred and in fact spread beyond the region that was in contact with the electrolyte. The cations therefore had to have been transported under the disbonded film. Calculation of ionic resistance under the disbonded coating, assuming an electrolyte with pH 14 and a disbonded gap of 1 μm, results in many orders of magnitude lower resistance than that measured for protective paint films [10]. Stratmann et al. observed that ions diffused along the interface from the defect, even when no disbonding had occurred [4]. The oxygen was removed from the electrolyte by argon and no disbonding was observed. However, the diffusion of ions from the defect was still detected with a scanning Kelvin probe (SKP). Many disbonding experiments have been performed in a humid atmosphere, where the electrolyte is only added to the coating holiday, and there, experiments also give CD [4,7,23–28]. In these experiments, the cations could only have been transported under the disbonded coating. Later, a disbonding experiment was performed on a semi-immersed sample, where the coating holiday was placed in the electrolyte surface, so that disbonding downward occurred under immersion conditions, while disbonding upward occurred in a humid atmosphere. For upward disbonding, transport of cations would only be possible under the coating, while for downward disbonding, cations could go both through the coating and under the disbonded coating. This experiment resulted in identical disbonding rates upward and downward, which must mean that cations are transported under the disbonded coating [29].

Oxygen and water have been assumed to be transported through the coating due to the relatively high solubility of these in the coating [4]. Experimental evidence has also verified this. By attaching an aluminum foil on the coating surface, it has been shown that the disbonding is almost completely stopped [9,10]. Since cations must be transported under the disbonded coating, we must then conclude that most of the oxygen required for the disbonding is transported through the coating.

12.1.4 Cathodic Disbonding Mechanism

A major breakthrough on the investigation of the CD mechanism was the introduction of the SKP in coatings research in the 1990s [30]. With the SKP, the electrochemical potential under a coating can be measured. A detailed explanation of the SKP and how it is utilized in coatings research is beyond the scope of this book, but several papers have been published on the topic [25,30]. The SKP consists of a vibrating needle that is scanned over the coating. Depending on the work function of the metal surface below, an AC is induced in a circuit. The work function has been shown to be proportional to the electrochemical potential of the surface. The measurement is parallel to the scanning vibrating electrode technique (SVET), but is performed in atmosphere and not immersed.

With the SKP the potential distribution under the coating can be measured. Figure 12.2 shows the potential distribution as a function of time and distance from the coating defect during a CD experiment [25]. The diagram shows three characteristic areas with respect to the potential. On the right-hand side in the diagram, there is a high and constant potential of about 0 mV versus SHE. The potential is assumed to be determined by the $Fe^{3+} + e^- = Fe^{2+}$ reaction. This has been shown to correspond with an adhering coating. The sudden potential step that gradually moves to the left in the diagram with time has been shown to correlate with the disbonding front. The potential gradient between the coating defect and the disbonding front is caused by the migration of cations from the defect under the disbonded coating, to neutralize the negative charge produced by the cathodic reaction. The resistance in the electrolyte under the disbonded coating causes an ohmic potential drop, and the potential gradually increases as we move farther away from the original coating defect.

As CD is caused by the oxygen reduction reaction under the coating, it is reasonable to assume that the rate of disbonding is proportional to the rate of the cathodic reaction, for a given coating. The rate of the oxygen reaction can be activation controlled

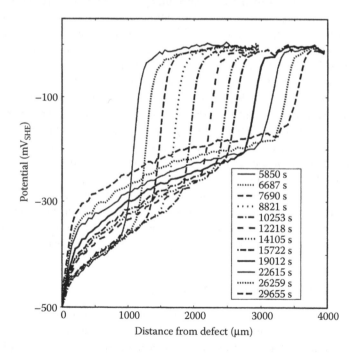

FIGURE 12.2 Potential distribution as a function of time and distance from the coating holiday under the coating during a CD experiment. (From Leng, A., et al., *Corros. Sci.*, 41, 547, 1999. Reprinted with permission.)

or diffusion limited. Under an organic coating, it may also be limited by transport of cations to the reaction site. The parameter studies summarized in Section 12.1.1 have shown that the disbonding rate is depending on cathodic potential, cation mobility, electrolyte concentration, and oxygen partial pressure at the same time. This should mean that the cathodic reaction is both activation controlled and diffusion limited at the same time, which apparently contradicts electrochemical theory.

The rate of CD has often been found to follow parabolic kinetics, which simply means that plotting disbonded distance against the square root of time gives a straight line:

$$x = k\sqrt{t} \tag{12.2}$$

where x is disbonded distance, k is a constant, and t is time. A plot of disbonded area against time will then of course also produce a straight line. This has been interpreted to mean that the disbonding rate is limited by diffusion of cations to the disbonding front, since Fick's law of diffusion will result in parabolic kinetics in this case [1]. However, the cations are not transported to the disbonding front by diffusion, but by migration. The cations are not moving due to a concentration gradient, but instead due to an electric field. The fact that there are no chlorides under the disbonded coating must be interpreted this way. Diffusion would result in transport of both anions and cations, while migration will be selective to cations. Fick's law of diffusion should then not be relevant, but we will arrive at the same parabolic law by deriving an equation for migration. Hence, if migration of cations is limiting the disbonding rate, we will observe parabolic kinetics. A general power law of disbonding kinetics, not assuming any specific delamination mechanism, can be expressed as [31]

$$x = k t^{a} \tag{12.3}$$

This equation can be transformed to

$$\log(x) = \log(k) + a \cdot \log(t) \tag{12.4}$$

According to Equation 12.4, plotting the logarithm of disbonded distance against the logarithm of time should produce a straight line. If the disbonding follows parabolic kinetics, then the slope of the line should be 0.5, corresponding to the square root of time. The alternative to transport limitation is activation control; that is, the disbonding rate is determined by the electrochemical potential. The slope of the line in case of pure activation control should be 1. Figure 12.3 shows the CD of an epoxy coating plotted against time, and plotted according to Equation 12.4. A line has been fitted to the logarithmic plot, and the equation is shown in the chart. In this case, the slope of the line was 0.79, that is, between what we would expect in the case of pure transport control and pure activation control. CD testing of a range of commercial and model epoxy coatings on blast-cleaned steel yielded slopes in the range

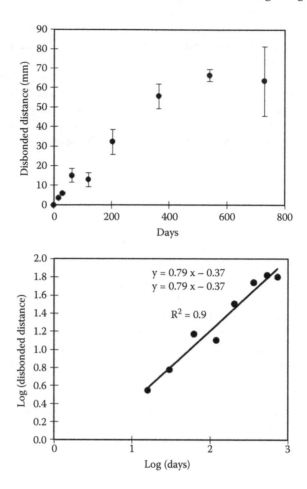

FIGURE 12.3   CD of a 300 μm thick epoxy coating on blast-cleaned steel.

of 0.7–0.9. Model coatings on polished steel and electrogalvanized steel have yielded slopes closer to 0.5 [31]. These plots show that the disbonding rate is not purely controlled by transport of cations to the disbonding front.

Simply exposing painted steel in a pH 14 solution of NaOH results in very little CD [8,32]. Evidently, the adhesion loss is not caused by diffusion of the hydroxide under the adhering coating. The hydroxide must be produced from cathodic oxygen reduction under the adhering coating immediately in front of the disbonding front. This agrees well with the mechanism by which organic coatings protect against corrosion, presented in Chapter 2. According to the mechanism, the organic coating stabilizes the oxide on the metal surface, and the oxide protects the metal from corrosion. The oxide separates the anodic reaction under the oxide and the cathodic reaction over the oxide. When the oxide has grown sufficiently thick, the resistance in the oxide prevents exchange of ions and electrons between the anodic and cathodic reactions, and the reactions stop. During cathodic disbonding, the cathode reaction on top of the oxide starts again, now balanced by cations supplied from the outside.

Hence, the cathodic reaction can go, but not the anodic reaction. We must then explain how the cations migrate under the adhering coating, before the disbonding occurs. Again, the SKP has provided interesting results that explain the observation. First, the steel–coating interface has been shown to be a preferential path for transport of ions [33,34], with a transport rate of up to five orders of magnitude higher compared with bulk polymer transport rates. Second, the steep potential gradient at the disbonding front, shown in Figure 12.2, is able to make cations migrate some distance along the steel–coating interface under the disbonded coating [26]. With ions at the steel–coating interface, we may actually have the conditions required for reduction of oxygen. We also have a low, but certain ionic conductivity that will connect the cathodic reaction under the adhering coating to the anodic reaction taking place externally, either in the coating holiday or on a connected anode.

This may also help us explain the apparently contradicting results that the disbonding rate is affected by both applied cathodic potential and oxygen concentration, indicating that the oxygen reaction should be activation controlled and diffusion controlled at the same time. Even though most coatings have a certain oxygen permeability, the oxygen flux through the coating has been shown to be low. With the rather low potentials measured under the disbonded coating (Figure 12.2), it is reasonable to assume that all the oxygen that comes through will react immediately upon reaching the metal surface, that is, diffusion control of the reaction rate. In the disbonding front, however, the potential is rapidly increasing. At some small distance ahead of the disbonding front, the potential will be so high that the oxygen reaction becomes activation controlled. A little farther away, the oxygen reaction will even stop, due to the lack of cations to neutralize the hydroxide. The lower the potential is at the disbonding front, the larger the potential difference we will have between the disbonding front and the potential under the unaffected coating. If we can assume a constant resistance against migration of cations under the adhering coating at the disbonding front, then a higher potential difference will mean a higher power for polarization under the coating. Thus, the cathodic reaction may take place farther in under the adhering coating. An oxygen concentration will result in a higher flux of oxygen through the coating, resulting in a higher disbonding rate, since the oxygen reaction is diffusion controlled. A lower potential will result in a higher potential difference at the disbonding front, allowing the oxygen reaction to occur farther in from the disbonding front. The documented effects of cation type and concentration will affect the resistance in the electrolyte under the disbonded coating, and thereby the potential drop between the coating damage and the disbonding front, that is, the potential at the disbonding front.

12.1.5 LIMITING CATHODIC DISBONDING

A few things can be done to limit the disbonding, but complete prevention is difficult to achieve. On the other hand, submerged steel that is protected by a combination of coatings and cathodic protection (CP) will still be protected by the anodes or passivated by the high pH when the coating is disbonded. Hence, as long the CP system is designed for a certain degradation of the coating, the structure will still be protected.

The following may decrease the rate of CD:

- Apply sufficient film thickness. The rate of CD decreases with increasing film thickness. For an epoxy paint system exposed in seawater, 350 μm is usually considered sufficient.
- Epoxy coatings with metallic aluminum pigments are less sensitive to CD. The aluminum pigments are chemically active during CD. The high pH caused by the cathodic reaction is aggressive to the aluminum pigments, so they corrode. In the corrosion process, some of the hydroxide is consumed, which slows down the disbonding. In order to have an effect, the aluminum pigments must be in the first coat [22].
- Avoid very low potentials in the CP system. The CD rate increases when the potential decreases. Zinc and aluminum anodes give cathodic potentials in the order of -1.1 $V_{Ag/AgCl}$. CP by impressed current may give lower potentials near the anode and higher disbonding rates. However, other aspects must also be taken into consideration when you are designing a CP system, so you are not always free to choose the optimum potential with respect to CD.

12.2   CORROSION CREEP

Most of us are familiar with this type of degradation and have seen this in action on coated steel. When a protective organic coating is damaged in some way, so that the metal is exposed to the environment, the bare metal starts to corrode. At first, the corrosion is limited to the exposed metal. However, after a relatively short time the corrosion starts to spread around the initial damage, degrading more and more of the coating. Figure 12.4 shows corrosion creep under the paint in a ballast water tank on a ship, but all painted steel is subject to this type of degradation. Corrosion creep may happen on all painted metals; in this chapter, we focus on steel since corrosion creep is the main degradation mechanism on painted steel

FIGURE 12.4   Corrosion creep initiating from welds and corners. (Photo: Børge Aune Schjelderupssen, Statoil.)

exposed in corrosive atmospheres. Corrosion creep on zinc (painted zinc coatings) is discussed in Chapter 13. Aluminum and magnesium are primarily attacked by filiform corrosion, which is addressed in Section 12.3.

Underfilm corrosion, undercutting corrosion, and anodic undermining are other names for this type of coating degradation.

12.2.1 INITIATION SITES FOR CORROSION CREEP

As mentioned above, at any site where the steel substrate is exposed, corrosion creep may start spreading. The exposed site may be mechanical damages in the paint from impacts, cracks, and so forth; paint application errors; or irregularities in the substrate.

Typical initiation sites are edges and welds. Figure 12.4 shows a picture from a ballast water tank after approximately 20 years in operation. From edges and welds, corrosion is spreading under the coating. There are few initiation sites on flat surfaces. The reason why corrosion mainly starts from edges and welds is that the coating usually is thinner there. All liquids will try to decrease their surface area due to their surface tension. Over sharp edges, the wet paint will reduce its surface area by withdrawing from the edge, and the paint will be a lot thinner than on the flat surfaces. Figure 12.5 shows a cross section of a coated edge. On the flat surfaces near the edge, the film thickness is on the order of 250 μm, while over the edge, it is less than 100 μm. Welds have irregular surface topography that creates the same problem. In fact, any irregularity in the surface will cause this problem, such as large blasting particles stuck in the surface or weld spatter. If possible, edges and welds should therefore be grinded before they are painted. Welds are grinded flat and edges are rounded to a radius of 2 mm. In addition, the edges and welds should be stripe coated, which means that every coat is applied in one extra layer by brush before the entire surface is spray-painted.

There are many application errors that potentially may lead to corrosion damages, and a full overview will not be given here. The most common errors are probably too-low film thickness and various types of pores. During paint application, it is almost impossible to have full control of the film thickness, in particular when the paint is applied manually. Robotic paint application, application by electrophoresis, and roller coating usually give a more uniform film thickness. Manual spray painting inevitably results in variation in film thickness. The film thickness is only

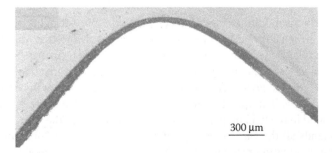

300 μm

FIGURE 12.5   The coating withdraws from sharp edges and corners due to surface tension, resulting in a thin coat that is prone to degradation.

controlled by spot checks, so on large constructions, areas with too-low film thickness are almost inevitable. This is one of the reasons why two thin coats of paint usually give a better result than one thick one. For statistical reasons, it is less likely that both coats will be too thin in the same area, so the two coats will increase the chance that the entire structure will have sufficient film thickness to be protected.

Pores is a term that is used for small holes in the paint film caused by a number of errors:

- Craters or fish-eyes are formed when the substrate is contaminated with oil or other substances with low surface energy that prevents the paint from wetting the substrate. The paint will withdraw from the contaminated area and leave a hole in the paint.
- Popping of the paint occurs when solvent evaporation causes formation of bubbles in the wet paint film. They may burst at a later stage when the wet paint has become so viscous that the film does not float together and close the hole [35].
- Pinholes are formed when the wet paint is applied over a porous substrate. The paint may not form a continuous film over these pores for various reasons, for example, insufficient paint is applied to fill the pore, or outburst of solvent or gas from the pore when the paint film is too viscous to close the hole, as in during popping (see above).

12.2.2 PROPAGATION MECHANISMS

The mechanism principally depends on properties of the substrate, the paint coating, and the environment where it is exposed. Different mechanisms have been suggested, but these three are the most recognized:

1. CD in the degradation front detaches the organic coating, exposing the fresh steel surface that subsequently is attacked by corrosion [36,37].
2. Anodic undermining—the metal corrodes away under the adhering coating [38].
3. Wedging—corrosion products lift the organic coating from the substrate around the attacked areas, exposing the fresh steel surface [39].

Which mechanism dominates will depend on type of coating, type of substrate, pretreatment of the substrate, and exposure conditions. More than one mechanism may also act at the same time, confusing the situation even further. Since CD will result in adhesion loss without corrosion of the substrate, this mechanism will show itself by an uncorroded area in front of the corrosion creep, when the detached coating is removed. This is frequently seen in various accelerated coating tests, but less frequently found in the field on painted steel constructions. In the field, the corrosion normally extends all the way to the degradation front, indicating that anodic undermining is the dominating mechanism. CD is a slow process on modern heavy-duty protective coatings, especially at free corrosion potentials, which also speaks for the anodic undermining mechanism.

In order for wedging to be the main propagation mechanism, the adhesion of the coating will have to be weaker than the cohesion of the coating. If it were not, the coating would tend to break rather than detach. For heavy-duty protective coatings applied on blast-cleaned steel, the adhesion is simply too strong for oxide wedging to occur.

As for CD studies, the SKP has proven to be a useful tool also for the study of corrosion creep. SKP measurement of electrochemical potential beneath an organic coating has to be performed in the atmosphere, that is, conditions that are normally associated with corrosion creep degradation. As shown in Section 12.1, when a thick layer of electrolyte covers the coating damage, the coating around the damage will degrade by CD. The coating damage acts as an anode and corrodes, while only the cathodic reaction will take place under the coating surrounding the coating damage. Hence, the coating fails by CD. The characteristic potential profile shown in Figure 12.2 results. However, if the electrolyte in the coating damage is allowed to dry, then the situation changes completely. The potential profile is inversed, and the steel in the coating damage becomes the cathode, while the steel under the coating surrounding the damage becomes the anode [38]. It is assumed that the potential in the damage increases due to increased transport of oxygen to the steel. With a shorter diffusion path through the thinner electrolyte film, the oxygen transport rate will increase. More oxygen will also push the Fe^{2+}/Fe^{3+} equilibrium toward Fe^{3+}, which may help keep the potential high if the electrolyte film thickens at a later stage. Reduction of Fe^{3+} back to Fe^{2+} may maintain the same potential gradient direction. During CD, sodium ions migrate in under the coating to balance the OH^- formed in the cathodic reaction, resulting in high pH and passivation of the steel. Now, when the anodic reaction is taking place under the coating, Cl^- will migrate in to balance the Fe^{2+} ions that are formed in the anodic reaction, as illustrated in Figure 12.6. Hydrolysis of the iron chloride under the coating may result in acidification of the electrolyte.

$$2FeCl_2 + 2H_2O = 2Fe(OH)_2 + 2HCl \qquad (12.5)$$

Due to the acidification of the electrolyte, hydrogen evolution may start, which will increase the corrosion rate further. The process resembles traditional crevice corrosion, although there are some important differences also. The size of the crevice between the organic coating and the corroding substrate will allow oxygen to

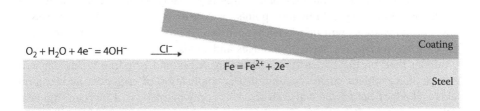

FIGURE 12.6 Schematic presentation of corrosion creep with acidification of the electrolyte under the organic coating. In a marine environment, Cl^- will be the dominating anion and migrate to the anode under the coating. Formation of iron chloride results in acidification of the electrolyte.

diffuse into the crevice. We also have the possibility of oxygen transport through the organic coatings into the crevice. Cathodic reduction of oxygen inside the crevice will result in formation of hydroxide and limit the acidification of the electrolyte.

12.2.3 LIMITING CORROSION CREEP

Many things can be done to limit corrosion creep, but most of them can be sorted into four categories:

1. Selecting a coating that is able to resist the environment in which it will be exposed
2. A paint-friendly design—the designer of the construction must take into account from the start that the structure will be painted
3. Performing a proper surface preparation and cleaning before the paint is applied
4. Quality control during the paint application process, ensuring that the paint is applied as specified

When selecting a coating, a sufficient barrier against ions is a prerequisite for a durable coating. As explained in Chapter 2, the coating protects the substrate primarily by creating a local environment on the metal surface where electrochemical reactions (corrosion) will not occur. When ions penetrate the coating, this is no longer true and corrosion will start [40]. The barrier properties are determined by both generic type and film thickness. Chemically curing paints are in general better barriers than physically drying paint. Applying a zinc-rich primer will also be helpful for atmospherically exposed steel, since it may provide some CP of the steel substrate. Zinc corrosion products may also create a less aggressive environment under the paint, since zinc chloride does not react with water to form hydrochloric acid, unlike iron and aluminum.

Paint-friendly design is about making sure that all surfaces can be accessed for cleaning and paint application. Avoiding typical corrosion traps, such as overlapping joints and design features that will trap water or dirt, will also decrease the chance for coating failure. ISO 12944-3 gives specific advice on design considerations for metal structures to be painted [41].

Insufficient surface preparation or cleaning is probably the most common reason for coating failure. Typical initiation points for corrosion creep have been discussed above; avoiding these will help prevent corrosion creep. As shown in Figure 12.4, corrosion creep typically starts from edges and welds, where the coating often will be thinner than on the flat surfaces. Rounding edges and grinding welds flat will decrease this problem. Stripe coating, that is, application of each coat by brush on welds and edges before the entire structure is spray-painted, will increase the film thickness on these features, further reducing the chance for initiation of corrosion. Proper cleaning to remove salt, grease, dirt, and so forth, is mandatory. Blast cleaning to a white metal finish and a good anchoring profile will ensure good adhesion and increase the lifetime of the coating. Blast cleaning and surface pretreatments are discussed in Chapter 9.

Finally, quality control during painting is required. At all construction sites, certified coating inspectors should inspect the work during and after surface preparation and painting. Surface cleanliness, blasting profile, film thickness, and so forth, must be controlled. Experience has shown that there is much truth to the saying "You get what you inspect, not what you specify." A good coating specification is a good starting point, but inspections are required to make sure that the coating is applied as specified. Several organizations are offering courses for coating inspectors, for example, NACE, FROSIO, and ICorr.

12.3   FILIFORM CORROSION

Filiform corrosion is threadlike corrosion attacks under the paint film, spreading from a coating damage. The term *filiform corrosion* stems from the word *filament*, which is derived from Latin and means "thread shaped." Filiform corrosion may occur on coated aluminum, steel, and magnesium, but is mainly associated with aluminum. An example of filiform corrosion on aluminum is shown in Figure 12.7. The picture shows a sample of coated aluminum after a filiform corrosion test. The sample was inoculated with diluted hydrochloric acid and put in a humidity cabinet at 40°C and 82% relative humidity (RH) for 1000 hours.

12.3.1   Filiform Corrosion Mechanism

Filiform corrosion occurs in a humid atmosphere, typically above 75% RH. The extent of attack will depend on the time the coated surface is wet [42], which is why surfaces in the shade that dry more slowly are more vulnerable. The problem is mainly found with rather thin lacquers and varnishes. The corrosion attacks are rather shallow, and filiform corrosion is therefore usually considered mainly to be an aesthetic problem. The anodic attack has been shown to occur in the filament head, while the cathodic reaction takes place in the tail behind [43,44]. The presence of chlorides is essential [45]. The filament head contains cations from the anodic

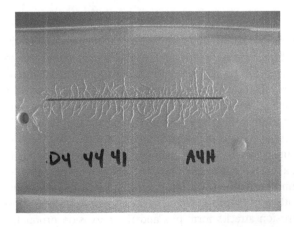

FIGURE 12.7   Filiform corrosion on aluminum.

reaction and chlorides. The reactants, that is, oxygen and water, are assumed to be transported in the tail [46]. The chlorides are transported with the propagating attack, and the corrosion products are precipitated as oxides or hydroxides in the tail as the pH increases due to the cathodic reaction.

As stated above, the anodic reaction is taking place in the filament front, while the cathodic reaction mainly is taking place in the tail. The role of the chlorides during filiform corrosion is to destabilize the metal oxide and acidification of the electrolyte. The growth of the filament is driven by a differential aeration cell between the anodic head and the cathodic tail. Schematically, the mechanism for filiform corrosion on aluminum can be described as follows [46]:

- At a weak point in the coating, for example, an edge or a mechanical damage, chlorides have access to the metal surface. The chlorides attack the metal oxide, as most passive metals are sensitive to corrosion when exposed to chlorides, and the metal starts to corrode.
- The anodic reaction moves under the organic coating, while the cathodic oxygen reduction takes place at the bare metal. The chlorides are transported to the anodic sites under the organic coating to maintain the charge balance.
- The metal cations hydrolyze in the chloride-containing electrolyte, resulting in formation of hydrochloric acid. The electrolyte is therefore acidified, and pH as low as 1–2 has been measured. This generally destabilizes the surface oxide and the substrate corrodes actively.
- The anodic reaction moves ahead, and the chlorides migrate after, to maintain the charge balance. The cathodic oxygen reduction takes place behind the anodic filament head, neutralizing the electrolyte and repassivating the tail. Hence, the acidified anodic filament head moves forward, while the oxygen reaction moves after.
- The hydrochloric acid is constantly regenerated by migration of the chlorides, along with the filament head.

12.3.2 FILIFORM CORROSION ON ALUMINUM

Filiform corrosion on aluminum received a lot of attention in the 1990s due to increased use of prepainted rolled aluminum façade panels in the architectural industry. Research projects were started in order to investigate the mechanism, after coating failure by filiform corrosion on a few high-profile buildings in northern Europe [47]. This led to the discovery of an activated and deformed surface layer that is formed during rolling and heat treatment of aluminum. The activated and deformed surface layer gave very rapid corrosion rates, and was a severe problem on coated aluminum panels. The mechanism for the degradation is briefly explained below [48–57], along with the mechanism in absence of this layer.

On aluminum, the filament growth may take place in two different ways, either by successive pitting or by the activated surface layer described above. In successive pitting, the corrosion attacks form pits and the corrosion products lift the organic coating from the substrate, exposing new metal surface where corrosion may initiate.

Coating adhesion will then be an important parameter for the filament growth. Low adhesion gives wide filaments and rapid filament growth, while good adhesion gives narrow filaments and slow growth. On aluminum with an activated surface layer, the filaments propagate by a different mechanism. Rolled aluminum alloys often have a layer with a deformed structure in the surface, typically as shown in Figure 12.8. To the right of the black line in the picture, the aluminum has a normal metallurgical microstructure, while the deformed microstructure shows to the right. The deformed layer is characterized by nanosized metal grains, rolled in oxide fragments and small intermetallic particles. During heat treatment of the metal at temperatures of >350°C, for example, for hardening, such layers are activated by diffusion of traces of lead or other group III or IV metals in the periodic table to the metal surface, destabilizing the protective oxide. Filiform corrosion then rapidly propagates in this layer, undercutting the organic coating. Since the problem is caused by the activated layer on the aluminum substrate, the propagation of the filaments is almost independent of the properties of the coating. Otherwise, the mechanism is similar to the one described above.

In the early 1990s, filiform corrosion posed a significant threat to the use of aluminum façade panels in the architectural industry. Rolled aluminum alloys are more susceptible due to the presence of deformed surface layers, which are not found on

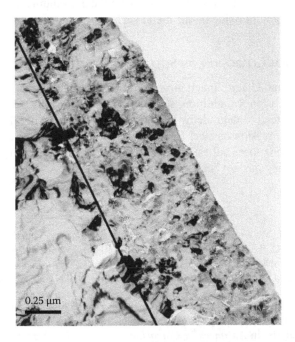

0.25 μm

FIGURE 12.8 Deformed surface layer on rolled aluminum. The deformed layer to the right of the black line is formed during rolling and is characterized by nanosized grains, rolled in aluminum oxide fragments and small intermetallic particles. The presence of such layers makes coated aluminum highly susceptible to filiform corrosion, particularly if the alloy has been heat treated to temperatures of >350°C before coating.

extruded or cast aluminum, or steel. Due to the important role of chlorides in the mechanism, the problem has mainly been found in coastal areas.

For rolled aluminum alloys, removal of activated and deformed surface layers is necessary. This is achieved by etching in either alkaline or acid electrolytes. Applying an effective conversion coating, passivating the aluminum, and providing a good basis for coating adhesion will also improve the filiform corrosion resistance. Chromating and anodizing are two types of pretreatment processes that give highly efficient conversion coatings in this respect. Due to the toxic properties of hexavalent chromium, chromating is being banned from more and more products and industries, and alternatives have been developed. The new chromate-free processes mainly function as adhesion promoters, contributing little or not at all to the passivation of the aluminum. Still, the effect on filiform corrosion may be significant, and good filiform corrosion resistance has been reported for many chromate-free processes [58–60].

The properties of the organic coating are also important for the filiform corrosion resistance of the coated product. Particularly good adhesion seems to be vital. In the aerospace industry, coating products pigmented with hexavalent chromium, for example, strontium chromate, have been extensively used historically. The presence of hexavalent chromium in the coating has an effect similar to that of chromate conversion coatings. When the coating is exposed to humid environments, hexavalent chromium leaches out of the coating, reacts with the aluminum substrate, and gives a passivating chromium oxide on the metal surface.

12.3.3 Filiform Corrosion on Steel

Corrosion creep and CD are much more common degradation mechanisms on steel than filiform corrosion. In addition, steel usually requires rather thick coatings for protection; these tend to degrade by corrosion creep or CD rather than filiform corrosion. Filiform corrosion on steel follows the general mechanism described above. Whether the coating loses adhesion in the filament head by wedging or cathodic or anodic undermining has not been established yet [44].

REFERENCES

1. Leidheiser, H., W. Wang, and L. Igetoft. *Prog. Org. Coat.* 11, 19, 1983.
2. Jin, X.H., K.C. Tsay, A. Elbasir, and J.D. Scantlebury. Adhesion and disbonding of chlorinated rubber on mild steel. In *Advances in Corrosion Protection by Organic Coatings*, ed. D. Scantlebury and M. Kendig. Pennington, NJ: Electrochemical Society, 37, 1987.
3. Kendig, M., R. Adisson, and S. Jeanjaquet. The mechanism of cathodic disbonding of hydroxy-terminated polybutadiene on steel from acoustic microscopy and surface energy analysis, in *Advances in Corrosion Protection by Organic Coatings*, ed. D. Scantlebury and M. Kendig. Pennington, NJ: Electrochemical Society, 1989.
4. Stratmann, M., R. Feser, and A. Leng. *Electrochim. Acta* 39, 1207, 1994.
5. McLeod, K., and J.M. Sykes. Blistering of paint coatings on steel in sea water, in *Coatings and Surface Treatment for Corrosion and Wear Resistance*, ed. K.N. Strafford, P.K. Datta, and C.G. Googan. Chichester: Ellis Horwood Ltd., 295, 1984.

6. Ritter, J.J. *J. Coat. Technol.* 54, 51, 1982.
7. Leng, A., H. Streckel, K. Hofmann, and M. Stratmann. *Corros. Sci.* 41, 599, 1999.
8. Steinsmo, U., and J.I. Skar. *CORROSION* 50, 934, 1994.
9. Sørensen, P.A., K. Dam-Johansen, C.E. Weinell, and S. Kiil. *Prog. Org. Coat.* 68, 283, 2010.
10. Knudsen, O.Ø., and J.I. Skar. Cathodic disbonding of epoxy coatings—Effect of test parameters. Presented at CORROSION/2008. Houston: NACE International, 2008, paper 08005.
11. Leidheiser, H., Jr., and W. Wang. *J. Coat. Technol.* 53, 77, 1981.
12. Watts, J.F., J.E. Castle, P.J. Mills, and S.A. Heinrich. Effect of solution composition on the interfacial chemistry of cathodic disbondment. In *Corrosion Protection by Organic Coatings*, ed. M.W. Kendig and H. Leidheiser. Pennington, NJ: Electrochemical Society, 1987.
13. Sykes, J.M., and Y. Xu. Investigation of electrochemical reactions beneath paint using a combination of methods. Presented at the 5th International Symposium on Advances in Corrosion Protection by Organic Coatings, Cambridge, UK, September 14–18, 137, 2010.
14. Leidheiser, H., and W. Wang. *J. Coat. Technol.* 53, 77, 1981.
15. Watts, J.F., and J.E. Castle. *J. Mater. Sci.* 19, 2259, 1984.
16. Sørensen, P.A., S. Kiil, K. Dam-Johansen, and C.E. Weinell. *J. Coat. Technol. Res.* 6, 135, 2009.
17. Watts, J.F. *J. Adhesion* 31, 73, 1989.
18. Sørensen, P.A., C.E. Weinell, K. Dam-Johansen, and S. Kiil. *J. Coat. Technol. Res.* 7, 773, 2010.
19. Ge, X., et al. *ACS Catalysis* 5, 4643, 2015.
20. Horner, M.R., and F.J. Boerio. *J. Adhesion* 2, 141, 1990.
21. Watts, J.F., and J.E. Castle. *J. Mater. Sci.* 19, 2259, 1984.
22. Knudsen, O.Ø., and U. Steinsmo. *J. Corros. Sci. Eng.* 2, 13, 1999.
23. Grundmeier, G., W. Schmidt, and M. Stratmann. *Electrochim. Acta* 45, 2515, 2000.
24. Leng, A., H. Streckel, and M. Stratmann. *Corros. Sci.* 41, 579, 1999.
25. Leng, A., H. Streckel, and M. Stratmann. *Corros. Sci.* 41, 547, 1998.
26. Posner, R., O. Ozcan, and G. Grundmeier. Water and ions at polymer/metal interfaces. In *Design of Adhesive Joints under Humid Conditions*, ed. L.F.M. d. Silva and C. Sato. Berlin: Springer, 21, 2013.
27. Rohwerder, M., E. Hornung, and M. Stratmann. *Electrochim. Acta* 48, 1, 2003.
28. Stratmann, M., W. Furbeth, G. Grundmeier, R. Losch, and C.R. Reinhartz. Corrosion inhibition by absorbed monolayers. In *Corrosion Mechanisms in Theory and Practice*, ed. P. Marcus and J. Oudar. New York: Marcel Dekker, 373, 1995.
29. Bi, H., and J. Sykes. *Corros. Sci.* 53, 3416, 2011.
30. Stratmann, M., H. Streckel, and R. Feser. *Corros. Sci.* 32, 467, 1991.
31. Doherty, M., and J.M. Sykes. *Corros. Sci.* 46, 1265, 2004.
32. Sørensen, P.A., S. Kiil, K. Dam-Johansen, and C.E. Weinell. *Prog. Org. Coat.* 64, 142, 2009.
33. Stratmann, M. *CORROSION* 61, 1115, 2005.
34. Wapner, K., M. Stratmann, and G. Grundmeier. *Electrochim. Acta* 51, 3303, 2006.
35. Knudsen, O.Ø., J.A. Hasselø, and G. Djuve. Coating systems with long lifetime—Paint on thermally sprayed zinc. Presented at CORROSION/2016. Houston: NACE International, 2016, paper 7383.
36. Funke, W. *J. Coat. Technol.* 55, 31, 1983.
37. Reddy, B., M. Doherty, and J. Stykes. *Electrochim. Acta* 49, 2965, 2004.
38. Nazarov, A., and D. Thierry. *CORROSION* 66, 0250041, 2010.
39. Dickie, R.A. *Prog. Org. Coat.* 25, 3, 1994.

40. Nguyen, T., J.B. Hubbard, and J.M. Pommersheim. *J. Coat. Technol.* 68, 45, 1996.
41. ISO 12944-3. Paints and varnishes—Corrosion protection of steel structures by protective paint systems. Part 3: Design considerations. Geneva: International Organization for Standardization, 1998.
42. Cambier, S.M., D. Verreault, and G.S. Frankel. *CORROSION* 70, 1219, 2014.
43. Schmidt, W., and M. Stratmann. *Corros. Sci.* 40, 1441, 1998.
44. Williams, G., and H.N. McMurray. *Electrochem. Commun.* 5, 871, 2003.
45. Koehler, E.L. *CORROSION* 33, 209, 1977.
46. Ruggeri, R.T., and T.R. Beck. *CORROSION* 39, 452, 1983.
47. Scamans, G.M., A. Afseth, G.E. Thompson, Y. Liu, and X.R. Zhou. *Mater. Sci. Forum* 519–521, 647, 2006.
48. Afseth, A., J.H. Nordlien, G.M. Scamans, and K. Nisancioglu. *Corros. Sci.* 43, 2359, 2001.
49. Afseth, A., J.H. Nordlien, G.M. Scamans, and K. Nisancioglu. *Corros. Sci.* 43, 2093, 2001.
50. Afseth, A., J.H. Nordlien, G.M. Scamans, and K. Nisancioglu. *Corros. Sci.* 44, 2491, 2002.
51. Afseth, A., J.H. Nordlien, G.M. Scamans, and K. Nisancioglu. *Corros. Sci.* 44, 2529, 2002.
52. Afseth, A., J.H. Nordlien, G.M. Scamans, and K. Nisancioglu. *Corros. Sci.* 44, 2543, 2002.
53. Afseth, A., J.H. Nordlien, G.M. Scamans, and K. Nisancioglu. *Corros. Sci.* 44, 145, 2002.
54. Leth-Olsen, H., A. Afseth, and K. Nisancioglu. *Corros. Sci.* 40, 1195, 1998.
55. Leth-Olsen, H., and K. Nisancioglu. *CORROSION* 53, 705, 1997.
56. Leth-Olsen, H., and K. Nisancioglu. *Corros. Sci.* 40, 1179, 1998.
57. Leth-Olsen, H., J.H. Nordlien, and K. Nisancioglu. *Corros. Sci.* 40, 2051, 1998.
58. Knudsen O.Ø., S. Rodahl, J.E. Lein, *ATB Metallurgie* 45, 26, 2006
59. Lunder, O., B. Olsen, and B. Nisancioglu. *Int. J. Adhesion Adhesives* 22, 143, 2002.
60. Lunder, O., et al. *Surf. Coat. Technol.* 184, 278, 2004.

13 Duplex Coatings: Organic Coatings in Combination with Metal Coatings

Combining metallic and organic coatings, called duplex coatings, has been shown to give very long lifetimes in corrosive environments. The use of such coatings dates back to the first half of the 1900s; from the 1950s, the use has increased steadily. Duplex coatings are now used extensively in various industries and constructions. Primarily, zinc has been used in duplex coatings, with great success. Aluminum coatings have also been used, but in some cases, this has resulted in very rapid and extensive coating failure. Both zinc-based and aluminum-based duplex coatings are discussed in this chapter.

13.1 ZINC-BASED DUPLEX COATINGS

13.1.1 Zinc Coatings

Zinc coatings have a long history in the corrosion protection of steel. This section gives a short introduction to metallic zinc coatings, as a background for the discussion of zinc-based duplex coatings. Corrosion properties of zinc and zinc coatings have been thoroughly reviewed by Zhang [1].

Zinc coatings are applied for their sacrificial protection of steel and low self-corrosion rate. The open-circuit potential of zinc is typically in the range of -1.0 to -1.05 V versus Ag/AgCl, well below the protection potential of steel. The zinc will therefore behave as a sacrificial anode to steel, and steel exposed at small defects in the zinc coating will be protected. In atmospheric exposure, the cathodic protection will only work over a limited range, due to the requirement for electrolytic contact between the anode and the cathode. The anode and the cathode will only be electrolytically connected through the thin electrolyte film on the metal surface, often with a limited conductivity, which means that the potential will increase rapidly as we move away from the zinc coating edge in a coating damage. Hence, the zinc coating is unable to protect large areas of exposed steel.

Zinc coatings can have very low corrosion rates, especially if the atmosphere's corrosivity is moderate; long lifetimes can also be achieved in corrosive atmospheres. The American Galvanizers Association has estimated the time to first maintenance for hot-dip galvanizing as a function of atmospheric corrosivity and coating thickness (Figure 13.1) [2]. The low corrosion rate of the zinc is due to the formation of

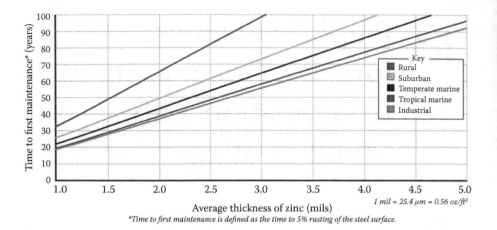

FIGURE 13.1 Time to first maintenance of hot-dip galvanizing in various atmospheric environments as a function of coating thickness. (From American Galvanizers Association, Zinc coatings: A comparative analysis of process and performance characteristics, American Galvanizers Association, Centennial, CO, 2011. Reprinted with permission.)

a zinc carbonate ($ZnCO_3$) film on the surface, passivating the zinc [1]. Zinc ions combine with carbonate from CO_2 in the air and the layer precipitates. Formation of a dense and protective carbonate layer requires that the surface dries out from time to time. This layer is vital for the longevity of the zinc coating, and when formation of the layer is prevented, very high corrosion rates can be found, on the order of 30–100 μm/year [1]. This is typically found in immersion conditions or under constant wetting. Very pure water and soft water are very aggressive to zinc coatings due to the low contents of carbonate, resulting in formation of nonprotective zinc hydroxide.

Industrially, the most important methods for applying metallic zinc coatings are hot-dip galvanizing, electrogalvanizing, and thermal spraying.

Hot-dip galvanizing is a multistep dipping process consisting of degreasing, pickling, fluxing, and finally the hot-dip galvanizing. The steel is fully immersed in molten zinc at 435°C–460°C, typically for about five minutes. The zinc coating is formed in an alloying reaction between iron and zinc, and the resulting coating consists of layered zinc–iron phases with decreasing content of iron. Only the outer layer is pure zinc. The metallic bonds between the coating and substrate ensure very strong adhesion. The thickness of the zinc coating depends primarily on the composition of the steel, particularly silicon content, and the thickness of the steel material to be coated [2]. Hence, extending the dipping time will not always result in a thicker coating. A typical coating thickness is 60–100 μm. Hot-dip galvanizing is used for steel items that can fit into the molten zinc pot; there is effectually a size limitation to the parts that can be coated this way. The size limitations that apply to hot-dip galvanizing also apply to powder coating, so the two processes fit well together and constitute an effective way of producing highly protective coatings [3].

Electrogalvanizing is a process where zinc is electroplated onto the steel. Prior to the electroplating, the steel is degreased and pickled, similar to the hot-dip galvanizing process. Electroplating is slower than hot-dip galvanizing, so electrogalvanizing typically gives thinner zinc coatings, in the order of 10–20 μm. There is no alloying reaction with the steel, so adhesion is also weaker. The appearance, however, is more decorative. Unless the electroplated zinc coating is painted, the objects are typically exposed indoors due to the limited film thickness. Electroplating, and subsequent painting, has been used for corrosion protection of car bodies on a large scale since the 1990s, and is probably the largest application of duplex coatings based on electroplated zinc.

Thermal spraying is a process where a material is melted by an energy source and blown to the substrate where it solidifies and forms a film. A wide range of metals, composites, and polymers can be thermally sprayed, and a wide range of spraying techniques have been developed. For spraying zinc, arc spray or flame spray equipment are used. Arc spraying has a larger capacity in terms of square meters per hour, while the equipment for flame spraying is smaller and easier to handle. Thermal spraying is the method for application of zinc on large constructions, such as steel bridges, offshore oil and gas platforms, and ships. The steel must be thoroughly blast cleaned before thermal spraying in order to achieve sufficient adhesion. The coating adheres to the steel by mechanical interlocking with the blasting profile. The coating is porous and has a very rough surface. A sealer coat must therefore be applied first, before thicker protective films can be applied.

13.1.2 LIFETIME OF ZINC-BASED DUPLEX COATINGS: SYNERGY EFFECT

Zinc-based duplex coating systems are extensively used in several industries, such as automotive, offshore wind energy, maritime, architecture, and infrastructure. The plated layer of zinc under the lacquer on most car bodies is the reason why corrosion is much less of a problem on cars today than it was in the 1980s. Offshore wind turbine towers, where coating maintenance is very difficult due to limited access for personnel, are often protected by thermally sprayed zinc (TSZ) and an organic coating system on top [4]. In many countries, road and rail bridges in coastal environments are also protected with TSZ and paint, due to their very long lifetime, typically 100 years.

Along the coast of Norway, there are more than 2000 steel bridges of different sizes and types of construction (steel beam, truss, suspension, ferry quay bridges, etc.). Many of these are exposed in a highly corrosive marine environment. Some floating bridges are even exposed in marine splash zones. The Norwegian Public Roads Administration (NPRA) estimates that they have about 2.5 million m² of coated steel on their bridges, and most of this area is coated with TSZ duplex coatings. They started to use TSZ duplex coatings in the 1960s, so they now have considerable experience with this type of coating. The use of duplex coatings is considered to be highly successful by the NPRA. Compared with bridges coated with conventional coating systems, the duplex-coated bridges generally have much longer maintenance intervals, and therefore, in spite of higher application costs, lower life cycle costs. One well-documented case is the Rombak Bridge outside Narvik (Figure 13.2), which is a suspension bridge crossing a fjord in northern Norway [5,6]. The bridge is 765 m long, the longest span is 325 m, and the sailing clearance is 41 m.

FIGURE 13.2   Rombak Bridge over the Rombak fjord. No maintenance of the coating system has been necessary since it was coated with TSZ duplex coating in 1970. (Photo: Reidar Klinge, NPRA.)

The bridge was opened for traffic in 1964. The bridge had then only received a temporary coating system, and not long after, corrosion initiated on much of the steel construction. In 1970, the bridge was coated on-site with a duplex coating system. The coating system applied was 100 μm TSZ, a wash primer, 80–100 μm of alkyd paint with zinc chromate, and 80–100 μm of alkyd finishing paint. The corrosivity on the truss work under the bridge has never been measured, but being located 41 m above the fjord, we can reasonably assume corrosion class 4 (ISO 12944-2). During an inspection of the bridge in 2009, after 39 years of service without maintenance, no corrosion was observed [6]. The topcoat had a rather dull appearance, so a new topcoat was applied for aesthetic reasons. Figure 13.3 shows rivet joints on the truss work under the bridge. Such overlapping joints are typical corrosion traps; corrosion often appears early at such joints. The lack of corrosion here therefore testifies to the protective properties and durability of the coating.

The durability of painted TSZ has also been documented by others. An extensive field test of TSZ with and without organic coatings was reported by the Naval Civil

FIGURE 13.3   Detail from Rombak Bridge. No corrosion can be found almost 40 years after application of the original coating. Even at lap joints like this, where the coating is unable to penetrate into the crevice between the overlapping steel parts, there is no sign of corrosion. (Photo: Reidar Klinge, NPRA.)

Engineering Laboratory in 1978 [7]. Ten-foot test panels with the various coating systems had been exposed in the tidal zone for 21 years at Port Hueneme in California. Some of the results from this test are summarized in Table 13.1. The samples were prepared in the 1950s, so the organic coatings are outdated, but the results shown in Table 13.1 effectively illustrate the protective properties of TSZ duplex coatings. The zinc coatings were applied by flame spray. The samples were exposed without scribe, and the coatings were evaluated with respect to the time red rust (iron oxides) started to penetrate the coating. Except for the vinyl acrylic, painting the zinc increased the lifetime of the coating by about four times. If applied directly on steel, the paint coatings used would probably last less than five years. The combination of a metallic zinc coating and paint gave longer lifetimes than the sum of the individual coats applied separately; that is, there was a synergy effect.

The synergy effect of zinc-based duplex coatings has been documented by Van Eijnsbergen [8]. He gathered experiences with hot-dip galvanized and painted steel in various corrosive environments around Europe, with lifetimes on the order of 15–25 years. Based on the findings from this study, and long-term field testing of similar duplex coatings, he proposed a model for the lifetime of the duplex coating system as a function of the lifetime of the zinc coating and the organic coating, expressed as

$$D_{\text{duplex}} = K \cdot \left(D_{\text{Zn}} + D_{\text{paint}} \right)$$

where D_{duplex} is the durability of the duplex coating system, D_{Zn} is the durability of the zinc coating alone, D_{paint} is the durability of the organic coating alone, and K is a synergy factor. He found synergy factors between 1.5 and 2.3, depending on

TABLE 13.1

Results from Port Hueneme Simulated Pile Test with TSZ Duplex Systems

TSZ	Thickness (μm)	Topcoat	Thickness (μm)	Time to failure (years)
Zn powder	75	—		2
Zn wire	140	—		4
Zn wire	100	—		4
Zn wire	60	Saran	180	18
Zn wire	60	Vinyl red lead	100	20
Zn wire	75	Vinyl acrylic	125	5
Zn wire	60	Vinyl finish	110	15
Zn wire	75	Epoxy finish	170	18
Zn wire	75	Gray furan	100	15

Source: Alumbough, R.L., and Curry, A.F., Protective coatings for steel piling: Additional data on harbor exposure of ten-foot simulated piling, Publication NCEL-TR-711S, final report, Naval Civil Engineering Laboratory, Port Hueneme, CA, 1978.

the corrosivity of the environment. The synergy factor increased when the corrosivity decreased. The synergy must be explained by differences in the degradation mechanism of the organic coating when it is applied on zinc instead of steel. This is discussed in the next section.

13.1.3 PROTECTION AND DEGRADATION MECHANISM FOR ZINC-BASED DUPLEX COATINGS

Zinc-based duplex coatings protect the steel by a combination of two mechanisms. Obviously, the zinc provides cathodic protection of steel exposed at small damages in the coating. The other protection mechanism is that the paint film displaces water from the zinc surface, passivating the zinc, as noted for steel in Chapter 2. Hence, the zinc protects the steel and the organic coating protects the zinc.

There are limitations to the capacity of the zinc coatings for cathodic protection. In humid atmospheres, zinc can only polarize the steel below the protection potential for a limited distance from the zinc edge. Depending on the properties of the water film formed on the metal surface, in particular the water film thickness and salt content, we can expect cathodic protection between a few millimeters and a few centimeters from the edge of the zinc. The cathodic reaction on the steel will produce hydroxide ions, while the anodic dissolution of zinc will produce zinc ions. Ions must migrate in the water film on the metal surface in order to maintain charge balance. The ionic current between the anode and cathode will result in an ohmic potential loss as a function of the distance from the zinc. Hence, at some distance from the zinc, the potential will exceed the protection potential and the steel is not protected anymore.

The synergy effect of zinc duplex coatings must depend on the durability of the organic coating on top of the zinc. Once the organic coating has failed, the zinc is exposed to the environment and its lifetime will be comparable to that of an unpainted zinc coating. This implies that corrosive degradation of the organic coating in zinc-based duplex coatings is slower than that for when the same organic coating is applied on steel.

When considering the degradation mechanism of duplex coatings by corrosion, we will have two quite different situations when a coating damage only penetrates the organic coating and exposes the zinc, compared with when the coating damage also penetrates the zinc, exposing the steel substrate. In the first case, only the zinc is exposed and the degradation is determined by the zinc and the organic coating alone. In the latter case, the steel will also have an effect. To complicate matters even further, the degradation may propagate by both cathodic disbonding and anodic undermining. In the following paragraphs, we investigate what happens in the various cases. Both the mechanism for cathodic disbonding of organic coatings on zinc [9–12] and anodic undermining [13,14] have been studied, although not as extensively as when they occur on steel.

Cathodic disbonding from a damage that exposed only zinc will take place in a manner similar to that described for steel in Chapter 12, although at lower potentials [12]. A galvanic cell is formed around the damage, where the anodic reaction takes place at the exposed zinc, and the cathodic oxygen reduction takes place under the coating surrounding the damage. Sodium ions migrate in under the coating to balance the hydroxide formed in the cathodic reaction, and an alkaline electrolyte is formed. However, contrary to steel, zinc is not passive at high pH, and a certain oxidation of the zinc will occur. Hence, an anodic front will follow the delamination front. Some of the hydroxide is consumed in the corrosion reaction, which means that the zinc will act as a buffer [12].

If the zinc exposed on the coating damage is only covered by a very thin layer of electrolyte, the zinc will passivate and the potentials in the galvanic cell described above will invert [13], parallel to what was described for anodic undermining on steel in Chapter 12. As the anodic reaction is now taking place under the organic coating, chloride ions will migrate in under the organic coating to balance the positively charged zinc ions, resulting in the formation of zinc chloride. However, contrary to aluminum chloride and iron chloride that hydrolyze and acidify the electrolyte, zinc chloride is stable. The solubility in water is high, but it will not hydrolyze and cause acidification. This is an important difference from aluminum and iron, and probably contributes significantly to the good performance of zinc-based duplex coatings.

When the coating damage penetrates the zinc as well as the organic coating, and the steel substrate is exposed, the situation becomes a bit more complicated. Corrosion at damaged sites in zinc-based duplex coatings will sooner or later reach this stage, regardless of whether the initial damage penetrated down to the steel. The zinc exposed at the damaged site will corrode over time; eventually, the zinc will be gone, exposing the steel, unless of course the organic coating is repaired in a timely manner. The situation with bare steel in the coating damage has also been investigated with a scanning Kelvin probe, and a potential profile under the coating surrounding a damage similar to the one shown in Figure 13.4 has been found [11].

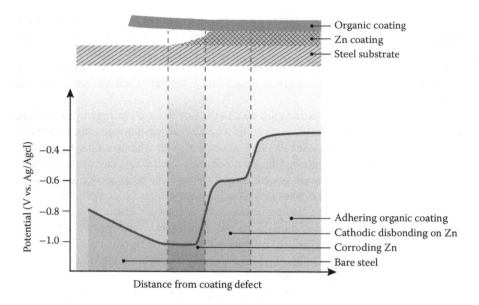

FIGURE 13.4 Potential profile under a degrading zinc-based duplex coating, where the steel substrate is exposed at the coating damage.

Around the coating damage, a corrosion cell is formed, where anodic dissolution of the zinc occurs under the paint. The corrosion of the zinc creeps under the organic coating, and with time, more and more steel is exposed. The corrosion of the zinc pulls down the potential on the exposed steel to a level where it is cathodically protected. Zinc corroding in a NaCl solution has a potential of about −1.0 V versus Ag/AgCl. A potential close to this, or somewhat higher due to polarization from the bare steel (mixed potential), can be expected here also. Cathodic oxygen reduction takes place on the steel, and exchange of ions between the anodic and cathodic sites occurs in the thin electrolyte on the metal surface. This ionic current causes a potential drop between the zinc anode and the steel cathode, and as we move farther away from the zinc along the bare steel surface, the potential gradually increases, parallel to what was described for a pure zinc coating earlier. At some distance from the zinc, the steel is no longer protected and it starts to corrode. In the opposite direction, in front of the corroding zinc, cathodic disbonding of the organic coating from the zinc has been found [11], marked by a potential plateau of about −0.6 V in Figure 13.4. In front of the cathodic disbanding, we find an even higher potential, representing the passive zinc under the intact protective organic coating.

Analysis of the metal surface by x-ray photoelectron spectroscopy (XPS) after removal of the disbonded organic coating has shown that the zinc ions formed in the anodic reaction are balanced by chloride ions that migrate in under the organic coating. As explained above, the benign nature of zinc chloride, not hydrolyzing and causing acidification of the local environment, probably explains to a large extent why zinc-based duplex coatings perform so much better than organic coatings applied directly on steel.

The situation described above is somewhat idealized and based on measurements performed in a laboratory on a model coating system in a stable, humid environment [11]. In the field, we will have cyclic drying and wetting, which most likely will affect the potential profile. The model coating system tested was also simpler and thinner than most coatings used in the field. Electrogalvanized zinc coatings will usually be phosphated before application of the organic coating, while hot-dip galvanized steel is normally sweep blasted or phosphated. This will increase adhesion of the organic coating and probably decrease the rate of cathodic oxygen reduction. These effects will slow down or eliminate the cathodic disbonding in front of the corroding zinc. In the field, thicker and better organic coatings will be used, which also will contribute to less cathodic disbonding. When investigating corrosion under zinc-based duplex coatings in field tests or cyclic aging tests (ISO 20340), cathodic disbonding is rarely found in front of the corroding zinc [3,15].

To summarize the mechanism for the beneficial effect of metallic zinc under a protective organic coating, we can say that

- The zinc provides cathodic protection of steel at sites where the coating is damaged, to a limited extent around the coating damage.
- To balance the zinc ions formed in the corrosion of the zinc, chloride ions migrate in under the organic coating and zinc chloride is formed. Opposite to steel, zinc chloride does not hydrolyze and the electrolyte under the coating is not acidified. Acidification would have enabled hydrogen evolution as a second cathodic reaction, accelerating the coating degradation. This is avoided with zinc-based duplex coatings.

13.1.4 KEY TO SUCCESS: ACHIEVING DURABLE ZINC-BASED DUPLEX COATINGS

As for all coatings, duplex coatings will also fail if they are not applied correctly. Errors during pretreatment of the steel, in application of the zinc coating, in pretreatment of the zinc, or in application of the organic coating may lead to poor coating quality and early failure.

On large constructions such as bridges and boats, the zinc coating must be applied by thermal spraying. TSZ coatings have a porous structure and a very rough surface, as shown in Figure 13.5, which is more challenging to spray-paint than conventional blast-cleaned steel. A sealer coat is required to fill the porous structure before thick organic coatings can be applied. The sealer is an organic coating and, in the case of TSZ duplex coatings, usually an epoxy that is diluted to a solid content of only 15%–40%. The high content of solvent gives a low viscosity, so that the sealer may penetrate the pores and complex geometry in the TSZ surface. The sealer should also be applied in a very thin coat, not building additional film thickness over the peaks in the TSZ, which means that the sealer will not hide the TSZ completely. You can still partly see the TSZ through the sealer. This way, the air in the pores easily escapes and the pores are filled by the sealer. Application of a too-viscous or too-thick sealer will result in pores in the coating, which may act as initiation points for corrosion [16]. Another potential problem with TSZ duplex coatings is so-called "spitting" during arc spraying. For various reasons, the metal may not be completely

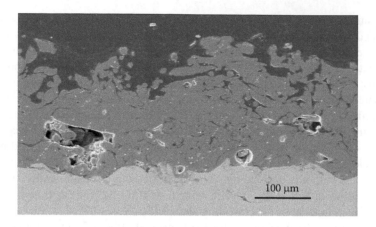

FIGURE 13.5   Cross section of TSZ coating applied on blast-cleaned steel.

melted in the spray gun and lumps of metal or even pieces of the spray metal wire may protrude from the produced metal coating. In a duplex coating, this will cause a weakness in the organic coating, because its film thickness may be too thin over the protrusion, serving as a site for initiation of corrosion. A trained metal sprayer will notice this when it happens and should take adequate corrective actions. The protrusions may be difficult to detect later on by visual inspection, but scraping a spattle over the metal coating may reveal them.

Electrogalvanized and hot-dip galvanized zinc coatings are often phosphated before the organic coating is applied, primarily to increase the adhesion of the organic coating to the zinc. The quality of the phosphate conversion coating is vital with respect to the coating quality [3]. A poor phosphating results in weak adhesion and more delamination around coating damages. Luckily, the quality of the phosphating is easily evaluated by visual inspection. Successful phosphating results in a matte, gray surface. Stroking a fingernail gently over the surface will result in a light gray track. If the zinc coating comes out of the phosphating with a shiny metal appearance, then the phosphating has been unsuccessful.

When applying an organic coating on hot-dip galvanized steel, there are a few issues to be aware of regarding the galvanizing. As explained in the introduction to this chapter, galvanizing is an alloying reaction, where the zinc reacts with the steel, forming various Zn-Fe phases, varying in Fe content. Figure 13.6 shows four cross sections of hot-dip galvanized coatings. The upper left picture shows a cross section of hot-dip galvanized low-Si steel (<0.03% Si + P). The various Zn-Fe phases can be seen, and in the outer surface there is a layer of pure zinc. These zinc coatings are usually easy to paint. The upper right picture shows dross or ash in the zinc coating surface. Dross is a by-product of the galvanizing process, where zinc in the kettle reacts with loose particles of iron and forms hard particles that may be trapped in the zinc coating. Ash is oxidized zinc floating on the surface of the zinc bath that may attach to the zinc coating when the steel is pulled out of the kettle. Over the protrusions, the organic coating will be thin, and these will then serve as initiation points for corrosion. The lower left picture shows hot-dip galvanized high-Si steel (>0.15% Si + P).

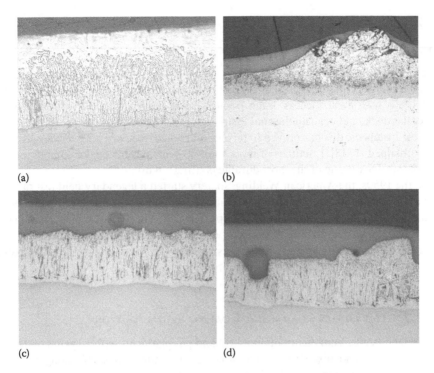

(a)   (b)

(c)   (d)

FIGURE 13.6   Hot-dip galvanized steel. (a) Zinc coating on low-Si steel. A layer of pure zinc in the outer surface of the zinc coating makes this zinc layer particularly suitable for application of organic coatings. (b) Dross in the zinc surface may penetrate the organic coating. (c) Galvanizing of high-Si steel gives Zn-Fe phases throughout the zinc coating. This may be challenging to paint. (d) Steel with 0.03%–0.15% Si is unsuitable for galvanizing and results in a crude zinc coating consisting of Si-Fe phases that is almost impossible to paint successfully.

The zinc coating formed on this type of steel may consist of Zn-Fe phases all the way to the surface. The surface then gets a rougher structure that may be more challenging to paint. The lower right picture shows the so-called Sandelin effect, which occurs when steel with a Si + P content of 0.03%–0.15% is hot-dip galvanized. The zinc coating grows in an uncontrolled manner and brittle Fe-Zn phases are formed [2]. The zinc coating has poor adhesion to the steel and the surface is difficult to paint.

Controlling the thickness of the organic coating can also be challenging in duplex coatings. Normally, film thickness is measured with a magnetic gauge, but since zinc is nonmagnetic, the magnetic gauge will measure the thickness of both the zinc and the organic coating. If the zinc coating is thicker than specified, then areas with too-low thickness in the organic coating are not discovered, since the total film thickness will be as specified. Thickness gauges based on eddy current will measure thickness of the organic coating alone, since the zinc coating is conductive. However, the conductivity of many zinc coatings varies locally; this will affect the measurements, yielding unreliable results. In TSZ coatings, for example, the porosity of the coating varies, affecting the coating conductivity.

13.2 ALUMINUM-BASED DUPLEX COATINGS

13.2.1 THERMALLY SPRAYED ALUMINUM

Thermally sprayed aluminum (TSA) has a long history as a durable and protective metal coating. In the offshore industry, it has been used on a large scale since the mid-1980s with great success. Flare booms, crane booms, lifeboat stations, below the cellar deck, and beneath thermal insulation are typical areas where TSA is used. In most instances, this has proven to be a good choice and very long service life has been obtained [17,18]. Lifetimes of more than 30 years have been documented, indicating that TSA is one of the most durable coatings in use.

In the 1950s, the American Welding Society started a legendary field test of TSZ and aluminum coatings in seawater and marine and industrial atmospheres. In 1974, results from 19 years of exposure were reported [19]. The TSA had performed well, and it was concluded that the lifetime expectancy was considerably longer than the duration of the test. It can be said that this marked the beginning of the widespread use of metallization for corrosion protection.

13.2.2 TSA DUPLEX COATINGS: A COATING SYSTEM TO AVOID

During the 1990s, a number of offshore oil and gas platforms with 50 years of design life were built and installed on the Norwegian continental shelf, along with an onshore gas treatment facility. With such long lifetimes, it was decided to use a stronger coating system than the regular three-coat paint systems consisting of a zinc-rich primer, an epoxy barrier coat, and a polyurethane topcoat. As discussed above, TSA was documented to give at least 30 years of service life in a marine environment. By applying a protective paint system on top of the TSA, an additional 20 years of service life was assumed, and 50 years of maintenance-free corrosion protection was expected. However, this turned out not to be the case. On the contrary, after a few years, massive coating degradation was found on all three offshore platforms. There was less degradation at the onshore gas treatment plant, but the same type of damage was found there too. The aluminum coating was corroding under the paint, and the corrosion was spreading at a high rate (Figure 13.7). The degradation typically started from corners, coating edges, and where stainless steel was mounted into the coated steel structure, such as cable ladders and pipe clamps.

The rapid failure of painted TSA had actually been observed earlier. In the same test referred to in Section 13.1.2 on TSZ duplex coatings, reported by the Naval Civil Engineering Laboratory in 1978, TSA with and without organic coatings was also tested [7]. Results from this test with TSA are summarized in Table 13.2. As for the TSZ duplex coatings, the organic coatings are outdated, but it is interesting to note that the TSA coatings without an organic coating all gave better performance than the duplex systems. For the TSZ coatings, the result was opposite—the bare TSZ failed and the TSZ duplex coatings performed well.

Another reporting of failed TSA duplex coatings came from the NPRA in 1975. Steel piles under Tromsøysund Bridge had been coated with a TSA duplex coating down to 1 m below the lowest tide. The organic coating started to blister almost

FIGURE 13.7   Corrosion of painted TSA. The TSA is corroding under the paint, causing the blistering of the paint.

TABLE 13.2
Results from Port Hueneme Simulated Pile Test with TSA Duplex Systems

Thermal Spray Coating	Thickness (μm)	Topcoat	Thickness (μm)	Time to Failure (years)
Al powder	110	—		21
Al wire	125	—		10
Al wire	100	—		18
Al wire	90	Saran	180	6
Al wire	110	Red lead	100	4
Al wire	60	Vinyl acrylic	125	4
Al wire	75	Vinyl finish	110	7

Source: Alumbough, R.L., and Curry, A.F., Protective coatings for steel piling: Additional data on harbor exposure of ten-foot simulated piling, Publication NCEL-TR-711S, final report, Naval Civil Engineering Laboratory, Port Hueneme, CA, 1978.

immediately after installation. Inside the blisters there was a liquid with pH 3.5–4. The TSA was corroded away inside the blisters and the steel surface was exposed.

After the failure of the TSA duplex coatings on the offshore platforms in the 1990s, an investigation of the degradation mechanism was started. The conclusion from this study was that TSA should not be covered by thick protective organic coatings [20]. A degradation mechanism was proposed where the TSA fails by anodic undermining. This is parallel to the mechanism for the corrosion of TSZ duplex coatings described above, but with one important difference: when TSA corrodes under the organic coating, aluminum chloride is formed instead of zinc chloride, which hydrolyzes and forms an acidic environment under the organic coating. The mechanism is illustrated in Figure 13.8.

Cathodic reduction of oxygen takes place at the bare steel, while anodic dissolution of the TSA takes place under the organic coating. In order to maintain charge balance, chloride ions migrate in under the organic to the corroding aluminum. The aluminum chloride hydrolyzes and the pH under the organic coating drops, probably below 4. Inside the blisters found on Tromsøysund Bridge, a liquid with pH 3.5–4 was found, and it is reasonable to assume that this generally is the case under failing TSA duplex coatings. The low pH in the electrolyte under the organic coating has two effects. First, the aluminum will corrode actively, since the protecting aluminum oxide is unstable at pH below 4. Second, hydrogen evolution from the acidic electrolyte will be an effective cathodic reaction. Hence, the corrosion of the aluminum does not depend on any external cathodic reaction, as we found in Section 13.2.1. The overall corrosion reaction will be

$$2Al + 6HCl \rightarrow 3H_2 + 2AlCl_3$$

In the corrosion reaction, aluminum chloride is formed again; that is, it will immediately react with water and regenerate the hydrochloric acid. As long as there is a supply of water, the corrosion reaction will maintain itself, which may explain the rapid failure found for TSA duplex coatings. The mechanism can be compared with crevice corrosion, where acidification of the electrolyte inside the crevice also plays an important role.

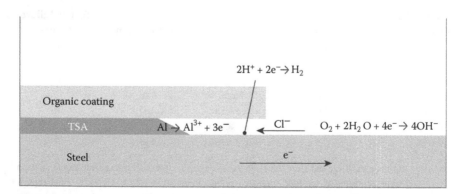

FIGURE 13.8   Corrosion of TSA duplex coatings in corrosive environments.

TSA coated only with a thin sealer has been shown to provide very good corrosion protection and little coating degradation, as shown in the introduction to this section. The explanation for this is simply that no crevice is formed as long as the sealer is applied correctly, that is, not building any additional film thickness and only filling the pores in the TSA.

REFERENCES

1. Zhang, X.G. *Corrosion and Electrochemistry of Zinc*. New York: Springer, 1996.
2. American Galvanizers Association. Zinc coatings: A comparative analysis of process and performance characteristics. Centennial, CO: American Galvanizers Association, 2011. Available at http://www.galvanizeit.org/ (accessed 2016).
3. Bjordal, M., S.B. Axelsen, and O.Ø. Knudsen. *Prog. Org. Coat.* 56, 68, 2006.
4. Mühlberg, K. *J. Prot. Coat. Linings* 21, 30, 2004.
5. Klinge, R. *Stahlbau* 68, 382, 1999.
6. Klinge, R. *Steel Construct.* 2, 109, 2009.
7. Alumbough, R.L., and A.F. Curry. Protective coatings for steel piling: Additional data on harbor exposure of ten-foot simulated piling. Publication NCEL-TR-711S, final report. Port Hueneme, CA: Naval Civil Engineering Laboratory, 1978.
8. Van Eijnsbergen, J.F.H. *Duplex Systems. Hot-Dip Galvanizing Plus Painting.* Amsterdam: Elsevier Science, 1994.
9. Fürbeth, W., and M. Stratmann. *Prog. Org. Coat.* 39, 23, 2000.
10. Fürbeth, W., and M. Stratmann. *Corros. Sci.* 43, 243, 2001.
11. Fürbeth, W., and M. Stratmann. *Corros. Sci.* 43, 229, 2001.
12. Fürbeth, W., and M. Stratmann. *Corros. Sci.* 43, 207, 2001.
13. Nazarov, A., M.G. Olivier, and D. Thierry. *Prog. Org. Coat.* 74, 356, 2012.
14. Ogle, K., S. Morel, and N. Meddahi. *Corros. Sci.* 47, 2034, 2005.
15. Knudsen, O.Ø., A. Bjorgum, and L.T. Dossland, *Mater. Perform.* 51, 54, 2012.
16. Knudsen, O.Ø., J.A. Hasselø, and G. Djuve. Coating systems with long lifetime—Paint on thermally sprayed zinc. Paper presented at CORROSION/2016. Houston: NACE International, 2016, paper 7383.
17. Døble, O., and G. Pryde. *Protect. Coat. Eur.* 2, 18, 1997.
18. Fischer, K.P., W.H. Thomason, T. Rosbrook, and J. Murali. *Mater. Perform.* 34, 27, 1995.
19. AWS (American Welding Society). Corrosion test of flames-sprayed coated steel: 19-year report. Miami: AWS, 1974.
20. Knudsen, O.Ø., T. Røssland, and T. Rogne. Rapid degradation of painted TSA. Paper presented at CORROSION/2004. Houston: NACE International, 2004, paper 04023.

14 Corrosion Testing: Background and Theoretical Considerations

The previous chapters have described various degradation processes acting on organic coatings in service, which lead to coating failure. Aging and breakdown of a good coating on a well-prepared substrate takes several years to happen in the field. Knowledge about the suitability of a particular coating is, of course, required on a much shorter time span; decisions about reformulating, recommending, purchasing, or applying a paint can often wait for a number of weeks or even a few months while test data are collected. Years, however, are out of the question. This explains the need for accelerated testing methods. The purpose of accelerated testing is to duplicate in the laboratory, as closely as possible, the aging of a coating in outdoor environments—but in a much shorter time.

This chapter considers testing the corrosion protection ability of coatings. A very brief explanation of some commonly used terms in corrosion testing of coatings is provided at the end of this chapter.

14.1 GOAL OF ACCELERATED TESTING

The goal of testing the corrosion protection ability of a coating is really to answer two separate questions:

1. Can the coating provide adequate corrosion protection?
2. What lifetime can be expected of the coating in the specific exposure environment?

The first question is simple: Is the coating any good at preventing corrosion? Does it have the barrier properties, or the inhibitive pigments, or the sacrificial pigments to ensure that the underlying metal does not corrode? The second question is, how will the coating hold up over time? Will it rapidly degrade and become useless? Or will it show resistance to the aging processes and provide corrosion protection for many years?

The difference may seem unimportant; however, there are advantages to separating the two questions. Testing a coating for initial corrosion protection is relatively inexpensive and straightforward. The stresses—water, heat, and electrolyte—that cause corrosion of the underlying metal are exaggerated, and then the metal under the coating is observed for corrosion. However, trying to replicate the aging process of a coating is expensive and difficult for several reasons:

1. Coatings of differing types cannot be expected to have a similar response to an accentuated stress.
2. Scaling down wet–dry cycles changes mass transport phenomena.
3. Climate variability means that the balance of stresses, and subsequent aging, is different from site to site.

The previous chapters have shown that organic coatings can fail in a number of different ways—chalking, blistering, loss of adhesion, cracking, CD, and corrosion creep, to mention the most common types of failure. Resistance against these failure modes depends on different coating properties, for example, adhesion, UV resistance, barrier, and electrochemical properties. The first property that fails will determine the lifetime of the coating. This has some implications for which properties should be tested and what exposure parameters should be accelerated.

14.2 ACCELERATED WEATHERING

The major weathering stresses that cause degradation of organic coatings are

- UV radiation
- Water and moisture
- Temperature
- Ions (salts such as sodium chloride [NaCl])
- Chemicals

The first of these weathering factors is unique to organic coatings; the latter four are also major causes of corrosion of bare metals. Most testing tries to reproduce natural weathering and accelerate it by accentuating these stresses. However, it is critically important to not overaccentuate them. To accelerate corrosion, we scale *up* temperature, salt loads, and frequency of wet–dry transitions; therefore, we must scale *down* the duration of each temperature–humidity step. The balances of mass transport phenomena, electrochemical processes, and the like necessarily change with every accentuation of a stress. The more we scale, the more we change the balances of transport and chemical processes from that seen in the field and the farther we step from real service performance. The more we force corrosion in the laboratory, the less able we are to accurately predict field performance.

For example, a common method of increasing the rate of corrosion testing is to increase the temperature. For certain coatings, the transport of ions increases markedly at elevated temperature. Even a relatively small increase in temperature above the service range results in large changes in the coating barrier properties. Such coatings are especially sensitive to artificially elevated temperatures in accelerated testing, which may never be seen in service. Other coatings, however, do not see strongly increased ion transport at the same elevated temperature. An accelerated test at elevated temperatures of these two coatings may falsely show that one was inferior to the other, when in reality both give excellent service for the intended application.

And, of course, interactions between stresses are to be expected. Some major interactions that the coatings tester should be aware of include

- *Frequency of temperature–humidity cycling.* Because the corrosion reaction depends on supplies of oxygen and water, the accelerated test must correctly mimic the mass transport phenomena that occur in the field. There is a limit to how much we can scale down the duration of a temperature–humidity cycle in order to fit more cycles in a 24-hour period. Beyond that limit, the mass transport occurring in the test no longer mirrors that seen in the field.
- *Temperature, salt load, and relative humidity (RH).* The balance of these factors helps to determine the size of the active corrosion cell. If that is not to scale in the accelerated test, the results can diverge greatly from that seen in actual field service. Ström and Ström [1] have described instances of this imbalance in which high salt loads combined with low temperatures led to an off-scale cell.
- *Type of pollutant and RH.* Salts such as NaCl and calcium chloride ($CaCl_2$) are hygroscopic but liquefy at different RHs. NaCl liquefies at 76% RH and $CaCl_2$ at 35%–40% RH (depending on temperature). At an intermediate RH, for example, 50% RH, the type of salt used can determine whether a thin film of moisture forms on the sample surface due to hygroscopic salts.

Various polymers, and therefore coating types, react differently to a change in one or more of these weathering stresses. Therefore, in order to predict the service life of a coating in a particular application, it is necessary to know not only the environment—average time of wetness (TOW), amounts of airborne contaminants, UV exposure, and so on—but also how these weathering stresses affect the particular polymer [2].

14.2.1 UV Exposure

UV exposure is extremely important in the aging and degradation of organic coatings. As the polymeric backbone of a coating is slowly broken down by UV light, the coating's barrier properties can be expected to worsen. However, UV exposure's importance in anticorrosion paints is strictly limited. This is because a coating can be protected from UV exposure simply by painting over it with another paint that does not transmit light.

The role of UV exposure in testing anticorrosion paints may be said to be "pass–fail." Knowing if the anticorrosion paint is sensitive to UV light is important. If it is, then it will be necessary to cover the paint with another coating to protect it from the UV light. This additional coating is routinely done in practice because the most important class of anticorrosion paints, epoxies, is notoriously sensitive to UV stress.

Because UV light itself plays no role in the corrosion process, the need for UV stress in an accelerated corrosion test is questionable. Besides, the sensitivity of the various coating types to UV is well known. Epoxies are very sensitive; polyurethanes, acrylics, polyesters, and alkyds are resistant; and polysiloxanes are even more resistant to UV.

Another question one can ask is whether chalking should be regarded as a coating failure. In applications where visual appearance is important, it definitely is. On, for example, unmanned offshore installations, chalking may be perfectly acceptable, if the coating remains protective.

14.2.2 Moisture

There are as many opinions about the proper amount of moisture to use in accelerated corrosion testing of paints as there are scientists in this field. The reason is almost certainly because the amount and form of moisture vary drastically from site to site. The global atmosphere, unless it is locally polluted (e.g., by volcanic activity or industrial facilities), is made up of the same gases everywhere: nitrogen, oxygen, carbon dioxide, and water vapor. Nitrogen and carbon dioxide do not affect coated metal. Oxygen and water vapor, however, cause aging of the coating and corrosion of the underlying metal. The amount of oxygen is more or less constant everywhere, but the amount of water vapor in the air is not. It varies depending on location, time of day, and season [3]. The form of water also varies: water vapor in the atmosphere is a gas, and rain or condensation is a liquid.

It is often noted that water vapor may have more effect on the coating than does liquid water. For nonporous materials, there is no theoretical difference between permeation of liquid water and that of water vapor [4]. Coatings, of course, are not solid, but rather contain a good deal of empty space, for example,

1. Pinholes are created during cure by escaping solvents or air.
2. Void spaces are created by crosslinking. As crosslinking occurs during cure, the polymer particles cease to move freely. The increasing restrictions on movement mean that the polymer molecules cannot be "packed" efficiently in the shrinking film. Curing is not a homogenous process, but proceeds by forming highly crosslinked microgels. When microgels meet toward the end of the curing, incomplete curing results due to lack of reactants and slow diffusion [5].
3. Void spaces are created when polymer molecules bond to a substrate. Before a paint is applied, polymer molecules are randomly disposed in the solvent. Once applied to the substrate, polar groups on the polymer molecule bond at reactive sites on the metal. Each bond created means reduced freedom of movement for the remaining polymer molecules. As more polar groups bond on reactive sites on the metal, the polymer chain segments between bonds loop upward above the surface (Figure 14.1). The looped segments occupy more volume and form voids at the surface, where water molecules can aggregate [6].

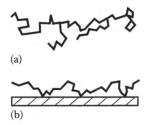

(a)

(b)

FIGURE 14.1   Looped polymer segments above the metal surface.

4. Spaces form between the binder and the pigment particles. Even under the best circumstances, areas arise on the surface of the pigment particle where the binder and the particle may be in extremely close physical proximity but are not chemically bonded. This area between the binder and pigment can be a potential route for water molecules to slip through the cured film.

Ström and Ström [1] have offered a definition of wetness that may be useful in weighing vapor versus liquid water. They have pointed out that NaCl liquidates at 76% RH, and $CaCl_2$ liquidates at 35%–40% RH (depending on temperature). NaCl is by far the most commonly used salt in corrosion testing. It seems reasonable to assume that unless the electrolyte spray–immersion–mist step in an accelerated test is followed by a rinse, a hygroscopic salt residue will exist on the sample surface. At conditions below condensing but above the liquidation point for NaCl, the hygroscopic residue can give rise to a thin film of moisture on the surface. Therefore, conditions at 76% RH or more should be regarded as wet. TOW for any test would thus be the amount of time in the cycle where the RH is at 76% or higher.

14.2.3   DRYING

A critical factor in accelerated testing is drying. Although commonly ignored, drying is as important as moisture. The temptation is to make the corrosion go faster by having as much wet time as possible (i.e., 100% wet). However, this approach poses three problems:

1. Studies indicate that corrosion progresses most rapidly during the transition period from wet to dry [7–11].
2. The corrosion mechanism of zinc in 100% wet conditions is different from that usually seen in actual service.
3. The degradation mechanism depends on RH. At high RH, the coating fails by cathodic disbonding (CD), while at low RH, it fails by anodic undermining [12].

14.2.3.1   Faster Corrosion during the Wet–Dry Transition

Stratmann and colleagues have shown that 80%–90% of atmospheric corrosion of iron occurs at the end of the drying cycle [8]; similar studies exist for carbon steel and zinc-coated steel. Ström and Ström [1] have reported that the effect of drying may be even more pronounced on zinc than on steel. Ito and colleagues [7] have provided convincing data of this as well. In their experiments, the drying time ratio, R_{dry}, was defined as the percentage of the time in each cycle during which the sample is subjected to low RH:

$$R_{dry} = \frac{T_{drying}}{T_{cycle}} \cdot 100\%$$

The drying condition was defined as 35°C and 60% RH; the wet condition was defined as 35°C and constant 5% NaCl spray (i.e., salt spray conditions). T_{cycle} is the total time, wet plus dry, of one cycle, and T_{drying} is the amount of time at 60% RH and 35°C during one cycle. Cold-rolled steel and galvanized steels with three zinc coating thicknesses were tested at R_{dry} = 0%, 50%, and 93.8%. For all four substrates, the highest amount of steel weight loss was seen at R_{dry} = 50%.

In summary, corrosion on both steel and zinc-coated steel substrates is slower if no drying occurs. This finding seems reasonable because, as the electrolyte layer becomes thinner while drying, the amount of oxygen transported to the metal surface increases, enabling more active corrosion [13–16]. As explained in Chapter 12, the wet–dry cycling changes the polarity between exposed steel in coating damages and steel under the surrounding coating. Under wet exposure, the coating damage is anode and the coated steel around is cathode, which means that the coating degrades by CD. When the humidity goes down, the polarity changes, and the coating degrades by anodic undermining [12].

Readers interested in a deeper understanding of this process may find the works of Suga [15] and Boocock [16] particularly helpful.

14.2.3.2   Zinc Corrosion—Atmospheric Exposure versus Wet Conditions

A drying cycle is an absolute must if zinc is involved either as pigment or as a coating on the substrate. The corrosion mechanism that zinc undergoes in constant humidity is quite different from that observed when there is a drying period. In field service, alternating wet and dry periods is the rule. Under these conditions, zinc can offer extremely good real-life corrosion protection—but this would never be seen in the laboratory if only constant wetness is used in the accelerated testing. This apparent contradiction is worth exploring in some depth.

Although this is a book about paints, not metallic corrosion, it becomes necessary at this point to devote some attention to the corrosion mechanisms of zinc in dry versus wet conditions. As discussed in Chapter 13, zinc-coated steel is an important material for corrosion prevention, and it is frequently painted (zinc-based duplex coating). Accelerated tests are therefore used on painted, zinc-coated steel. In order to obtain any useful information from accelerated testing, it is necessary to understand the chemistry of zinc in dry and wet conditions.

In normal atmospheric conditions, zinc reacts with oxygen to form a thin oxide layer. This oxide layer in turn reacts with water in the air to form zinc hydroxide ($Zn[OH]_2$), which in turn reacts with carbon dioxide in air to form a layer of basic zinc carbonate [17–19]. Zinc carbonate serves as a passive layer, effectively protecting the zinc underneath from further reaction with water and reducing the amount of corrosion. The degradation rate of painted zinc has been shown to depend on the CO_2 concentration in the atmosphere [20].

When zinc-coated steel is painted and then scribed to the steel, the galvanic properties of the zinc–steel system determine whether, and how much, corrosion will take place under the coating. The mechanism was explained in Chapter 13.

When Ito and colleagues [7] repeated their experiments with R_{dry} on painted, cold-rolled, and galvanized steels, an interesting pattern emerged. In Figure 14.2, the natural logarithm of the length of underfilm corrosion D is plotted against the R_{dry} for different zinc coating thicknesses. The relationship between zinc coating thickness, drying ratio, and underfilm corrosion distance is fairly distinct when presented thus.

Under wet conditions (i.e., low R_{dry}), more underfilm corrosion is seen on zinc-coated steel than on cold-rolled steel. Under dry conditions (i.e., high R_{dry}), the ranking is opposite and better results are obtained with zinc.

1. Zinc dissolves anodically at the front end of corrosion.
2. In the blister area behind the front end of corrosion, zinc at the top of the zinc layer dissolves due to OH, which is generated by a cathodic reaction.

However, if conditions include high R_{dry}, then underfilm corrosion is less on galvanized steel than on cold-rolled steel, for the following reasons:

1. The total supply of water and chloride (Cl⁻) is reduced, limiting cell size at the front end and zinc anodic dissolution area.

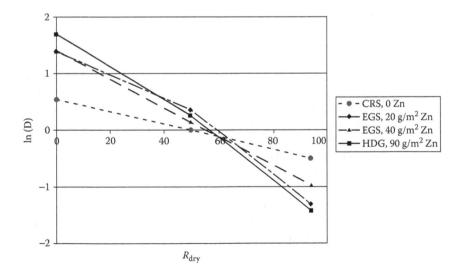

FIGURE 14.2 Natural log of underfilm corrosion, as a function of drying ratio for cold-rolled (CRS), electrogalvanized (EGS, 20 g/m² Zn and 40 g/m² Zn), and hot-dipped galvanized (HDG, 90 g/m² Zn) steel. (Data from Ito, Y., et al., *Iron Steel J.*, 77, 280, 1991.)

2. The electrochemical cell at the scribe is reduced.
3. Zinc is isolated from the wet corrosive environment fairly early. A protective film can form on zinc in a dry atmosphere. The rate of zinc corrosion is suppressed in further cycling.
4. The zinc anodic dissolution rate is reduced because the Cl⁻ concentration at the front end is suppressed.

It should be noted that the 90 g/m² zinc coating in this study is hot-dipped galvanized, and the two thinner coatings are electrogalvanized. It may be that differences other than zinc thickness—for example, structure and morphology of the zinc coating—play a not yet understood role. Further research is needed in this area, to understand the role played by zinc layer structure and morphology in undercutting.

14.2.3.3 Differences in Absorption and Desorption Rates

The rate at which a coating absorbs water is not necessarily the same as the rate at which it dries out. Some coatings have nearly the same absorption and desorption rates, whereas others show slower drying than wetting, or vice versa.

In constant stress testing, in which samples are always wet or always dry, this difference does not become a factor. However, as soon as wet–dry cycles are introduced, the implications of a difference between absorption and desorption rates become highly important. Two coatings with roughly similar absorption rates can have vastly different desorption rates. The duration of wet and dry periods in modern accelerated tests is measured in hours, not days, and it is quite possible that for a coating with a slower desorption rate, the drying time in each cycle is shorter than the time needed by the coating for complete desorption. In such cases, the coating that desorbs more slowly than it absorbs can accumulate water.

The problem is not academic. Lindqvist [21] has studied absorption and desorption rates for epoxy, chlorinated rubber, linseed oil, and alkyd binders, using a cycle of six hours of wet followed by six hours of drying. An epoxy coating took up 100% of its possible water content in the wet periods but never dried out in the drying periods. Conversely, a linseed oil coating in this study never reached its full saturation during the six-hour wet periods but dried out completely during the drying periods.

Lindqvist has pointed out that the difference in the absorption and desorption rates of a single paint, or of different types of paint, could go far in explaining why cyclic accelerated tests often do not produce the same ranking of coatings as does field exposure. There is a certain risk to subjecting different types of coatings with unknown absorption and desorption characteristics to a cyclic wet–dry accelerated regime. The risk is that the accelerated test will produce a different ranking from that seen in reality. It could perhaps be reduced by some preliminary measurements of water uptake and desorption; an accelerated test can then be chosen with both wet times and drying times long enough to let all the paints completely absorb and desorb.

14.2.4 TEMPERATURE

Temperature is a crucial variable in any accelerated corrosion testing. Higher temperature means more energy available, and thus faster rates, for the chemical processes that cause both corrosion and degradation of cured films. Increasing the temperature—within limits—does not alter the corrosion reaction at the metal surface; it merely speeds it up. A potential problem, however, is what the higher temperature does to the binder. If the chemical processes that cause aging of the binder were simply speeded up without being altered, elevated temperature would pose no problem. But this is not always the case.

Every coating is formulated to maintain a stable film over a certain temperature range. If that range is exceeded, the coating can undergo transformations that would not occur under natural conditions [3]. The glass transition temperature (T_g) of the polymer naturally limits the amount of acceleration that can be forced by increasing heat stress. Testing in the vicinity of the T_g changes the properties of the coatings too much, so that the paint being tested is not very much like the paint that will be used in the field—even if it came from the same can of paint.

14.2.5 CHEMICAL STRESS

When the term *chemical stress* is used in accelerated testing, it usually means chloride-containing salts in solution, because airborne contaminants are believed to play a very minor role in paint aging. See Chapter 11 for information about airborne contaminants.

Testers may be tempted to force quicker corrosion testing by increasing the amount of chemical stress. Steel that corrodes in a 0.05% NaCl solution will corrode even more quickly in 5% NaCl solution; the same is true for zinc-coated steel. The problem is that the amount of acceleration is different for the two metals. An increase in NaCl content has a much more marked effect for zinc-coated substrates than for carbon steel substrates. Ström and Ström [1] have demonstrated this effect in a test of weakly accelerated outdoor exposure of painted zinc-coated and carbon steel samples. In this weakly accelerated test, commonly known as the "Volvo scab" test, samples are exposed outdoors and sprayed twice a week with a salt solution. Table 14.1 gives the results after one year of this test, using different levels of NaCl for the twice-weekly spray.

From this study, it can be seen that raising the chloride load has a much stronger effect on painted zinc-coated substrates than on painted carbon steel substrates. It is known that for bare metals, the zinc corrosion rate is more directly dependent than the carbon steel corrosion rate on the amount of pollutant (NaCl in this case). This relationship may be the cause of the results in Table 14.1. In addition, higher salt levels leave a heavier hygroscopic residue on the samples (see Section 14.2.3); this may have caused a thicker moisture film at RH levels above 76%.

Boocock [22] reports another problem with high NaCl levels in accelerated tests: high saponification reactions, which are not seen in the actual service, can occur at high NaCl loads. Coatings that give good service in actual field exposures can wrongly fail an accelerated test with a 5% NaCl load.

Increasing the level of NaCl increases the rate of corrosion of painted samples, but the amount of acceleration is not the same for different substrates. As the NaCl load is increased, the range of substrates or coatings that can be compared with

TABLE 14.1

Average Creep from the Scribe after One Year of Weakly Accelerated Field Exposure

	Outdoor Samples Sprayed Twice per Week With		
Material	0.5% NaCl (mm)	1.5% NaCl (mm)	5% NaCl (mm)
Mean for all electrogalvanized and hot-dipped galvanized painted samples	1.3	2.0	3.1
Mean for all cold-rolled steel painted samples	6.2	8.2	9.6

Source: Modified from Ström, M., and Ström, G., A statistically designed study of atmospheric corrosion simulating automotive field conditions under laboratory conditions, SAE Technical Paper Series, paper 932338, Society of Automotive Engineers, Warrendale, PA, 1993.

each other in the test must narrow. A low salt load is recommended for maximum reliability.

Another approach is to reduce the frequency of salt stress. Most cyclic tests call for salt stress between two and seven times per week. Smith [23], however, has developed a cyclic test for the automotive industry that uses five-minute immersion in 5% NaCl once every two weeks. The high salt load—typical for when the test was developed—is offset by the low frequency.

How much salt is too much? There is no consensus about this, but several agree that the 5% NaCl used in the famous salt spray test is too high for painted samples. Some workers suggest that 1% NaCl should be a natural limit. Some of the suggested electrolyte solutions at lower salt loads (using water as solvent) are

- 0.05% (wt) NaCl and 0.35% ammonium sulfate, $(NH_4)_2SO_4$ [24]
- 0.5% NaCl + 0.1% $CaCl_2$ + 0.075% $NaHCO_3$ [25]
- 0.9% NaCl + 0.1% $CaCl_2$ + 0.25% $NaHCO_3$ [26]

14.2.6 ABRASION AND OTHER MECHANICAL STRESSES

While in service, coatings undergo external mechanical stresses, such as

- Abrasion (also called *sliding wear*)
- Fretting wear
- Scratching wear
- Flexing
- Impingement or impact

These stresses are not of major importance in corrosion testing. Even though some damage to the coating is usually needed to start corrosion, such as a scribe down to the metal, the formation of the coating damage is rarely included in the corrosion test. Instead, the test is started with a machined scribe in the coating. This is not to say that the area is unimportant: a feature of good anticorrosion coatings is that they can contain the amount of corrosion by not allowing corrosion creep to spread far from the original point of damage. Mechanical stress may be viewed in a manner similar to that for UV exposure: depending on the service application, it can be a pass–fail type of test. For example, in applications that will be exposed to a lot of stone chipping (i.e., because of proximity to a highway), impact testing may be needed.

There are several excellent reviews of external mechanical stresses, including details of their causes, their effects on various coating types, and the test methods used to measure a coating's resistance to them. For more information, the reader is directed toward several existing publications [27–29].

14.2.7 Implications for Accelerated Testing

Traditionally, accelerated testing of organic coatings has been attempted in the laboratory by exaggerating the stresses (heat, moisture, UV, and salt exposure) that age the coating. The prevailing philosophy has been that more stress equals more acceleration.

The previous sections have discussed why this prevailing philosophy is flawed. Based on this discussion, some limitations on stresses are proposed:

- Temperatures cannot be elevated above or anywhere near the T_g of the polymer.
- Moisture is important, but a drying cycle is equally important.
- Salt levels should be lower than those commonly used today.
- UV exposure is usually not necessary.

14.3 WHY THERE IS NO SINGLE PERFECT WEATHERING TEST

A great deal of research has gone into understanding the aging process of coatings, and attempts to replicate it more accurately and quickly in laboratories. Great advances have been made in the field, and even more advances are expected in the future. Still, we will never see one perfect accelerated test that can be used to predict coating performance anywhere in the world, on all coating types and all substrates. An accelerated test can never be used to predict service lifetime. It is impossible to convert the result of an accelerated test to degradation rate in the field. The best we can hope for is a certain correlation in ranking. The best coatings in a test can be expected to also be among the best in the field.

There are several reasons for this:

- Different sites around the world have different climates, stresses, and aging mechanisms.
- Different coatings have different weaknesses, and will not respond identically to an accentuated stress in the laboratory.

- It is not possible to accentuate all weathering factors and still maintain the balance between them that exists in the field.

These are discussed in more detail in the following sections.

14.3.1 DIFFERENT SITES INDUCE DIFFERENT AGING MECHANISMS

Sites can differ dramatically in weather. Take, for example, a bridge connecting Prince Edward Island to the Canadian mainland and a bridge connecting the island of Öland to the Swedish mainland. At first glance, one could say that these two sites are roughly comparable. Both are bridges standing in the sea, located closer to the North Pole than to the equator. Yet, these two sites induce different stresses in paints. A coating used on the first bridge would undergo much higher mechanical stress, due to heavy floes of sea ice. It would also see much higher salt loads because the Atlantic Ocean has a higher salt concentration than does the Baltic Sea. If these two sites, which at first glance seemed similar, can induce some differences in aging mechanisms, then the difference must be even more drastic between such coastal sites as Sydney, Vladivostok, and Rotterdam or between inland sites such as Aix-en-Provence, Brasilia, and Cincinnati.

The point is not academic; it is crucially important for choosing accelerated tests. A mechanically tough coating that is not particularly susceptible to salt would perform well at both sites, but an equally mechanically tough coating that allows some slight chloride permeation may fail at Prince Edward Island and succeed at Öland.

A study of coated panels exposed throughout pulp and paper mills in Sweden by Rendahl and Forsgren [30] illustrates the classic problem of using accelerated tests to predict coating performance: the ranking of identical samples can change from site to site. In this study, 23 coating–substrate combinations were exposed at 12 sites in 2 pulp and paper mills for five years. The sites with the most corrosion were the roofs of a digester house and a bleach plant at the sulfate mill. Although these two locations had similar characteristics—same temperature, humidity, and UV exposure—they produced different rankings of coated samples. Both locations agree on the worst sample, but little else. An alkyd paint that gave good results on the bleach plant roof had abysmally poor results on the other roof. Conversely, an acrylic that had significant undercutting on the bleach plant roof performed well on the digester house roof.

These results illustrate why there is no "magic bullet": an accelerated test that correctly predicts the ranking of the 23 samples at the digester house roof may be wildly wrong in predicting the ranking of the same samples at the bleach plant roof of the same mill.

Glueckert [31] has reported the same phenomenon based on a study of gloss loss of six coating systems exposed at both Colton, California, and East Chicago, Indiana. The East Chicago location had an inland climate, with a temperature range of −23°C to 38°C. The Colton site had a higher temperature, more intense sunlight, and blowing sand. The loss of gloss and ranking of the six coatings are shown in Table 14.2. The two sites identified the same best and worst coatings, but ranked the four in between differently.

TABLE 14.2

Exposure Results from Colton, California, and East Chicago, Indiana

Coating	Gloss Loss (%), East Chicago	Gloss Loss (%), Colton	Ranking, East Chicago	Ranking, Colton
Epoxy-urethane	3	0	1	1
Urethane	38	31	2	3
Waterborne alkyd	56	6	3	2
Epoxy B	65	83	4	5
Acrylic alkyd	68	77	5	4
Epoxy A	98	98	6	6

Source: Data from Glueckert, A.J., Correlation of accelerated test to outdoor exposure for railcar exterior coatings, presented at CORROSION/1994, NACE International, Houston, 1994, paper 596.

Another study of coatings exposed at various field stations throughout Sweden [2] found no correlations between sites in the corrosion performances of the identical samples, either in the amount of corrosion or in the ranking at each site. In this study, identifying a coating as "always best" or "always worst" was not possible.

Even if only one coating and one substrate were to be tested, it would not be possible to design an accelerated test that would perfectly suit all the exposure sites mentioned in this section—much less all the sites in the world.

14.3.2 DIFFERENT COATINGS HAVE DIFFERENT WEAKNESSES

Cured coatings are commonly thought of as simple structures: the usual depiction is a layer of binder containing pigment particles. The general view is that of a homogenous, continuous, solid binder film reinforced with pigment particles. In reality, the cured coating is a much more complex structure.

For one thing, instead of being a solid, it contains lots of empty space: pinholes, voids after crosslinking, gaps between pigment and binder, and so on. All these voids are potential routes for water molecules to slip through the cured film. What is important for accelerated testing is that the amount of empty space in the coating is not constant—it can change during weathering, as both the binder and the pigment change. Some pigments, such as passivating pigments, are slowly consumed, causing the empty space between the pigment and binder to increase. Other pigments immediately corrode on their surface. The increased volume of the corrosion products can decrease the empty space between particles and binder.

Binders also change with time, for many reasons. The stresses in the binder caused by film formation can be increased, or relieved, during aging. The magnitude of the stresses caused by film formation, and what happens to these stresses upon weathering, depends to a large extent on the type of polymer used for the binder. The same could be said for UV degradation, or any stress that ages binders: the binder's reaction, both in mechanism and in magnitude, depends to a large extent on the specific polymer used. Even if only one exposure site were to be used, it would not be possible to design an accelerated test that would be suitable for all binders and pigments.

14.3.3 STRESSING THE ACHILLES' HEEL

Every coating has its own Achilles' heel—that is, a point of weakness. The ideal test would accelerate all stresses to the same extent. It would then be possible to compare coatings with different aging mechanisms—different Achilles' heels—to each other.

Unfortunately, it is not possible to accentuate all stresses evenly. Furthermore, it is not possible to accentuate all weathering factors and still maintain the balance between them that exists in the field. When we increase the percentage of time with UV load, for example, we change the ratio of light and dark and move a step away from the real diurnal cycle seen in the field.

Because it is not possible to evenly accelerate all aging factors, the best testing tries to imitate an expected failure mechanism. Each test accentuates one or a few stresses that are rate controlling for a mechanism. By choosing the right test, it is possible to thus probe for certain expected weaknesses in the coating–substrate system. The trick, of course, is to correctly estimate the failure mechanism for a particular application, and thus pick the most suitable test.

14.4 ACCELERATED IMMERSION TESTING

Testing coatings for immersion service is in many ways simpler than for atmospheric exposure because the exposure environment varies less. Coatings for atmospheric exposure can experience temperatures from −50°C in the arctic to +50°C in a desert. Humidity can vary from completely dry to completely wet. There are environments completely free from salts, while in marine environments, coatings can be covered by salt deposits. The span in exposure conditions for immersed coatings will in most cases be much smaller:

- The temperature will typically be somewhere between 0°C and 30°C.
- The salinity varies from freshwater to seawater, but for a given construction it is normally given. The exception will be ships that go from the ocean and upriver.
- There is no UV radiation.

A new parameter is introduced, though—electrochemical potential. The potential of painted construction steel can vary from about −0.6 V to −1.5 V versus Ag/AgCl, depending on whether it is cathodically protected (CP).

The oxygen concentration will also vary in immersion service, which rarely is the case in atmospheric exposure. When buried in soil, which also is considered as a type of immersion, the oxygen concentration may be zero. Anaerobic conditions will also be found in subsea mud, tanks and pipelines, and some waters. In deep oceans, the oxygen concentration is lower than in shallow waters [32].

Both electrochemical potential and oxygen concentration may therefore be used to accelerate testing, in addition to temperature.

CD is considered the most important degradation mechanism for immersed coatings. Coatings for immersion service are therefore tested for CD. The effect of exposure parameters on CD was discussed in Section 12.1.

14.4.1 ELECTROCHEMICAL POTENTIAL

The tests are mainly accelerated by applying a lower potential than in service. CP structures in service typically have potentials between -0.8 and -1.1 V versus Ag/AgCl, while many tests use a potential of about -1.5 V. Below about -1.0 V, the dominating cathodic reaction is hydrogen evolution, while oxygen reduction dominates above. This may introduce a slight change in the degradation, since we may get some hydrogen evolution under the disbonded coating, especially early in the test, when there still is little disbonding. As it was shown in Section 12.1, the cathodic potential in the disbonding front is higher than the applied potential due to resistance under the disbonded coating. In the beginning of the test, when the disbonded distance is short, hydrogen may be possible. Later, when the disbonded distance is longer, the potential in the disbonding front will be too high for hydrogen evolution. This has, however, resulted in some skepticism to tests with low potentials, and, for example, ISO 15711 is not accelerated with respect to applied potential and uses -1.05 V versus Ag/AgCl, which is within the expected potential range for CP steel in seawater.

There are tests that use potentials as low as -3 V versus $Cu/CuSO_4$ in order to get results in a short time. Whether the results from such aggressively accelerated tests are reliable can be questioned.

14.4.2 OXYGEN CONCENTRATION

As shown in Section 12.1, the oxygen concentration in the electrolyte will affect CD. The disbonding test could therefore be accelerated by, for example, performing the test in a pure oxygen atmosphere, instead of air. So far, no test has utilized this, probably because it will be practically difficult to control the oxygen concentration in the electrolyte. Since these tests typically last from 30 days to 6 months, the amount of oxygen required would also be high, increasing the costs of the test.

The oxygen concentration in the test electrolyte will drop with time, unless measures are taken, usually by bubbling air through the electrolyte.

14.4.3 TEMPERATURE

Temperature will accelerate CD, but most tests are performed at ambient laboratory temperature, or expected service temperature if the temperature is elevated.

The pipeline industry has tested for CD at elevated temperatures for a long time [33–35], and most of the elevated temperature tests have been developed for this purpose. For structural steel, high-temperature CD has not been focused on until recently, when a test was included in NORSOK M-501 [36].

There has been some debate about whether the test vessel should be pressurized when testing at temperatures above 100°C. If not, the electrolyte may evaporate inside the disbonding crevice under the coating, which may slow down or stop the disbonding [37]. A thorough investigation of the subject has not been performed, but there are indications for an effect [37,38]. For deepwater constructions, the question is relevant, since the hydrostatic pressure will prevent the electrolyte from boiling. Buried onshore pipelines will be exposed at ambient atmosphere pressure, and pressurizing the test should not be necessary.

14.4.4   ELECTROLYTE COMPOSITION

As also shown in Section 12.1, the CD rate depends on the type of cation in the electrolyte. The disbonding rate increased with the cation type in the following order: $CaCl_2$, LiCl, NaCl, KCl, and CsCl. This order agrees with the mobility of the cations in water. Hence, performing the test in CsCl would also accelerate the test. No test utilizes this, even though the degradation mechanism would not be altered.

One aspect of CD testing related to electrolyte composition that needs some consideration is the fact that the test will alter the electrolyte. Polarizing the samples with a sacrificial anode will result in corrosion of the anode and release of ions into the electrolyte. Zinc, aluminum, or magnesium anodes can be used, and the ions from the anode may affect the test. Regular replacement of the electrolyte regularly or separating the anode from the test electrolyte via a salt bridge may be required. When polarizing the samples with a potentiostat or a galvanostat, an inert anode (counterelectrode) is used, but hypochlorite will be formed at the anode. Testing at ambient temperature has not shown any effect of hypochlorite in the test electrolyte [39]. However, at elevated temperature the hypochlorite will attack epoxy [33]. Again, isolating the test electrode via salt bridge or replacing the electrolyte regularly will solve the problem.

14.4.5   RELIABILITY OF CD TESTING

As explained above, the exposure environment for immersed coatings is more defined and predictable. This is probably why correlation between accelerated testing and testing under field conditions has been quite good [40]. Round-robin tests have shown fair correlation between labs performing the same CD test on parallel samples [41]. In the same round-robin study, aging tests for atmospheric coatings were also included, which gave a lot more variation.

14.4.6   RELEVANCE OF CD TESTING

The purpose with painting steel that will be CP is to decrease the current demand and reduce the amount of anodes. Hence, the coating is not providing corrosion protection. However, the CP design depends on the coating. Some CP design codes only specify coatings according to thickness and require no qualification testing [42], while others require CD testing for qualification [43].

An investigation of coating degradation on submerged steel constructions with CP after 20 years or more in service showed that the coatings mainly were in excellent condition [44]. Some blistering was found in a few areas, but the coatings were in place and covering the steel structure. The condition of the coatings was far better than anticipated by the CP design codes. There must have been CD spreading from coating edges and the blisters, but in spite of this, the coating was in place, performing its purpose of decreasing cathodic protection current demand. Laboratory testing has shown that there is no correlation between CD and current demand for CP [40]. Hence, one may ask whether CD tests are necessary. The answer to the question so far is that CD testing is a cheap guarantee that the coating has a certain quality. When the coatings perform well, the CP will last longer. Retrofitting anodes is very expensive and therefore unwanted. Besides, lifetime extension of offshore installations is rather the rule than the exception. High-quality coatings and long-life CP are desirable for this reason too.

CD testing will not only measure CD resistance, but also resistance against cathodic blistering. The test consists of direct current polarization of the coated sample, which will measure the barrier properties of the coating. If cations penetrate the coating, the cathodic reaction will start at the steel–coating interface, producing NaOH and subsequent alkaline blistering. Blistering will increase current demand for CP, so sufficiently good barrier properties are vital [40].

In high-temperature CD testing, the barrier aspect of the test is extra relevant. If the coating is degraded by the temperature, its barrier properties will deteriorate. This will be revealed by cathodic blistering.

REFERENCES

1. Ström, M., and G. Ström. A statistically designed study of atmospheric corrosion simulating automotive field conditions under laboratory conditions. SAE Technical Paper Series, paper 932338. Warrendale, PA: Society of Automotive Engineers, 1993.
2. Forsgren, A., and C. Appelgren. Performance of organic coatings at various field stations after 5 years' exposure. Report 2001:5D. Stockholm: Swedish Corrosion Institute, 2001.
3. Appleman, B. *J. Coat. Technol.* 62, 57, 1990.
4. Hulden, M., and C.M. Hansen. *Prog. Org. Coat.* 13, 171, 1985.
5. Nguyen, T., et al. *J. Coat. Technol.* 68, 45, 1996.
6. Kumins, C.A., et al. *Prog. Org. Coat.* 28, 17, 1996.
7. Ito, Y., et al. *Iron Steel J.* 77, 280, 1991.
8. Stratmann, M., et al. *Corros. Sci.* 27, 905, 1987.
9. Miyoshi, Y., et al. Corrosion behavior of electrophoretically coated cold rolled, galvanized and galvannealed steel sheet for automobiles—Adaptability of cataphoretic primer to zinc plated steel. SAE Technical Paper Series, paper 820334. Warrendale, PA: Society of Automotive Engineers, 1982.
10. Nakgawa, T., et al. *Mater. Process* 1, 1653, 1988.
11. Brady, R., et al. Effects of cyclic test variables on the corrosion resistance of automotive sheet steels. SAE Technical Paper Series, paper 892567. Warrendale, PA: Society of Automotive Engineers, 1989.
12. Nazarov, A., and D. Thierry. *CORROSION* 66, 0250041, 2010.
13. Mansfield, F. Atmospheric corrosion. In *Encyclopedia of Materials Science and Engineering*. Vol. 1. Oxford: Pergamon Press, 1986, p. 233.

14. Boelen, B., et al. *Corros. Sci.* 34, 1923, 1993.
15. Suga, S. *Prod. Finish* 40, 26, 1987.
16. Boocock, S.K. *J. Prot. Coat. Linings* 11, 64, 1994.
17. Seré, P.R. *J. Scanning Microsc.* 19, 244, 1997.
18. Odnevall, I., and C. Leygraf. Atmospheric corrosion. In ASTM STP 1239, ed. W.W. Kirk and H.H. Lawson. Philadelphia: American Society for Testing and Materials, 1994, p. 215.
19. Almeida, E.M., et al. An electrochemical and exposure study of zinc rich coatings. In *Proceedings of Advances in Corrosion Protection by Organic Coatings*, ed. D. Scantlebury and M. Kendig. Vol. 89-13. Pennington, NJ: Electrochemical Society, 1989, p. 486.
20. Fürbeth, W., and M. Stratmann. *Corros. Sci.* 43, 243, 2001.
21. Lindqvist, S. *CORROSION* 41, 69, 1985.
22. Boocock, S.K. Some results from new accelerated testing of coatings. Presented at CORROSION/1992. Houston: NACE International, 1992, paper 468.
23. Smith, A.G. *Polym. Mater. Sci. Eng.* 58, 417, 1988.
24. Mallon, K. Accelerated test program utilizing a cyclical test method and analysis of methods to correlate with field testing. Presented at CORROSION/1992. Houston: NACE International, 1992, paper 331.
25. Townsend, H. Development of an improved laboratory corrosion test by the automotive and steel industries. Presented at the 4th Annual ESD Advanced Coatings Conference. Detroit: Engineering Society of Detroit, 1994.
26. Yau, Y.-H., et al. Performance of organic/metallic composite coated sheet steels in accelerated cyclic corrosion tests. Presented at CORROSION/1995. Houston: NACE International, 1995, paper 396.
27. Hare, C.H. *J. Prot. Coat. Linings* 14, 67, 1997.
28. Koleske, J.V. *Paint and Coating Manual: 14th Edition of the Gardner-Sward Handbook.* Philadelphia: ASTM, 1995.
29. Paul, S. *Surface Coatings Science and Technology.* Chichester: John Wiley & Sons, 1996.
30. Rendahl, B., and A. Forsgren. Field testing of anticorrosion paints at sulphate and sulphite mills. Report 1997:6E. Stockholm: Swedish Corrosion Institute, 1998.
31. Glueckert, A.J. Correlation of accelerated test to outdoor exposure for railcar exterior coatings. Presented at CORROSION/1994. Houston: NACE International, 1994, paper 596.
32. Sverdrup, H.U., et al. *The Oceans, Their Physics, Chemistry, and General Biology.* New York: Prentice-Hall, 1942.
33. Kehr, J.A. *Fusion-Bonded Epoxy (FBE): A Foundation for Pipeline Corrosion Protection.* Houston: NACE International, 2003.
34. Mitschke, H.R., and P.R. Nichols. Testing of external pipeline coatings for high temperature service. Presented at CORROSION/2006. Houston: NACE International, 2006, paper 06040.
35. Surkein, M., et al. Corrosion protection program for high temperature subsea pipeline. Presented at CORROSION/2001. Houston: NACE International, 2001, paper 01500.
36. NORSOK M-501. Surface preparation and protective coatings. Rev. 6. Oslo: Norwegian Technology Standards Institution, 2012.
37. Knudsen, O.Ø., et al. Test method for studying cathodic disbonding at high temperature. Presented at CORROSION/2010. Houston: NACE International, 2010, paper 10007.
38. Cameron, K., et al. Critical evaluation of international cathodic disbondment test methods. Presented at CORROSION/2005. Houston: NACE International, 2005, paper 5029.

39. Knudsen, O.Ø., and J.I. Skar. Cathodic disbonding of epoxy coatings—Effect of test parameters. Presented at CORROSION/2008. Houston: NACE International, 2008, paper 8005.
40. Knudsen, O.Ø., and U. Steinsmo. Current demand for cathodic protection of coated steel—5 years data. Presented at CORROSION/2001. Houston: NACE International, 2001, paper 01512.
41. Winter, M. Laboratory test methods for offshore coatings—A review of a round robin study. Presented at CORROSION/2011. Houston: NACE International, 2011, paper 11041.
42. RP B401. Cathodic protection design. Oslo: Det Norske Veritas, 2005.
43. NORSOK M-503. Cathodic protection. Oslo: Norwegian Technology Standards Institution, 2016.
44. Knudsen, O.Ø., and S. Olsen. Use of coatings in combination with cathodic protection—Evaluation of coating degradation on offshore installations after 20+ years. Presented at CORROSION/2015. Houston: NACE International, 2015, paper 5537.

15 Corrosion Testing: Practice

This chapter provides information about

- Accelerated tests that age coatings
- Why the salt spray test should not be used
- What to look for after an aging test is completed
- Accelerated testing of coatings for immersion service
- Advanced methods for studying coatings and degradation
- How the amount of acceleration in a test is calculated, and how the test is correlated to field data

A large number of tests are available, but only a few are discussed here. Methods for structural steel in corrosive environments are the focus.

15.1 ACCELERATED AGING METHODS AND CORROSION TESTS

The aim of the accelerated test regime is to age the coating in a short time in the same manner as would occur over several years of field service. These tests can provide direct evidence of coating failure, including creep from the scribe, blistering, and rust intensity. They also are a necessary tool for the measurement of coating properties that can show indirect evidence of coating failure. A substantial decrease in adhesion or significantly increased water uptake, even in the absence of rust-through or undercutting, is an indication of imminent coating failure.

Hundreds of test methods are used to accelerate the aging of coatings. Several of them are widely used, such as salt spray and ultraviolet (UV) weathering. A review of all the corrosion tests used for paints, or even the major cyclic tests, is beyond the scope of this chapter. It is also unnecessary because this work has been presented elsewhere; the reviews of Goldie [1], Appleman [2], and Skerry and colleagues [3] are particularly helpful.

The aim of this section is to provide the reader with an overview of a select group of accelerated aging methods that can be used to meet most needs:

- General corrosion tests—all-purpose tests
- Condensation or humidity tests
- Weathering tests (UV exposure)

In addition, some of the tests used in the automotive industry are described. These are tests with proven correlation to field service for car and truck paints, which may, with adaptations, prove useful in heavier protective coatings.

A general accelerated test useful in predicting performance for all types of coatings, in all types of service applications, is the "holy grail" of coating testing. No test is there yet, and none probably ever will be (see Chapter 14). However, some general corrosion tests can still be used to derive useful data about coating performance. The two all-purpose tests recommended here are the ASTM D5894 test and the ISO 20340 test.

15.1.1   ISO 20340 (AND NORSOK M-501)

ISO 20340 actually contains test specifications for coatings to be exposed in a marine atmosphere, splash zone, and immersed zone. For an immersed zone, it refers to ISO 15711, which is discussed in Section 15.3.1. The aging cycle for atmospheric coatings is discussed here.

The test was developed for the offshore oil and gas industry, particularly the conditions found in the North Sea. Initially, the test was described in NORSOK M-501, but in 2003, the test became ISO 20340. That is also when the freeze–thaw step was included in the cycle. The NORSOK M-501 coating specification standard now refers to ISO 20340 for prequalification of coatings for atmospherically exposed steel below 120°C.

NORSOK M-501 provides requirements for materials selection, surface preparation, paint application, inspection, and so on, for coatings used on offshore platforms. However, other industries are also referring to NORSOK M-501 now. Due to the requirements in NORSOK M-501, a large number of coatings have been qualified according to ISO 20340, from which other industries now can benefit.

The aging cycle is 168 hours long, and it runs for 25 cycles (i.e., 25 weeks total). Each cycle consists of 72 hours of UV condensation (ASTM G53), 72 hours of salt spray, and 24 hours of freezing at 20°C. Between 1995 and 2000, a large number of coatings were tested in an offshore field test, and corrosion creep performance was correlated to the aging test [4]. Correlations between 0.6 and 0.75 were found.

Since the introduction of NORSOK M-501 in 1994, there have been three major coating failure histories on the Norwegian continental shelf [5]. The three cases are (1) painting of thermally sprayed aluminum (TSA), (2) topcoat flaking of two-coat paint films, and (3) cracking of a rapidly curing coating system. The first case was discussed in Chapter 13. In two of the three cases, the coatings actually did not pass the qualification test, but were still accepted with the deviation. Hence, following the prequalification requirements more rigorously would have prevented these incidents.

15.1.2   ASTM D5894 (AND NACE TM0404)

ASTM D5894, "Standard Practice for Cyclic Salt Fog/UV Exposure of Painted Metal (Alternating Exposures in a Fog/Dry Cabinet and a UV/Condensation Cabinet)," is also called "modified Prohesion" or "Prohesion UV." This test, incidentally, is sometimes mistakenly referred to as "Prohesion testing." However, the Prohesion test does not include a UV stress; it is simply a cyclic salt fog (one-hour salt spray, with 0.35% ammonium sulfate and 0.05% sodium chloride [NaCl], at 23°C, alternating with one drying cycle at 35°C). The confusion no doubt arises because the original developers of ASTM D5894 referred to it as "modified Prohesion."

NACE TM0404 refers to ASTM D5894 for aging and rust creep resistance testing.

This test can be used to investigate both anticorrosion and weathering characteristics. The test's cycle is 2 weeks long and typically runs for six cycles (i.e., 12 weeks total). During the first week of each cycle, samples are in a UV–condensation chamber for four hours of UV light at 60°C, alternating with four hours of condensation at 50°C. During the second week of the cycle, samples are moved to a salt spray chamber, where they undergo one hour of salt spray (0.05% NaCl + 0.35% ammonium sulfate, pH 5.0–5.4) at 24°C, alternating with one hour of drying at 35°C.

The literature contains warnings about too-rapid corrosion of zinc in this test; therefore, it should not be used for comparing zinc and nonzinc coatings. If zinc and nonzinc coatings must be compared, an alternate (i.e., nonsulfate) electrolyte can be substituted under the guidelines of the standard. This avoids the problems caused by the solubility of zinc sulfate corrosion products. It has also been noted that the ammonium sulfate in the ASTM D5894 electrolyte has a pH of approximately 5; at this pH, zinc reacts at a significantly higher rate than at neutral pH levels. The zinc is unable to form the zinc oxide and carbonates that give it long-term protection.

15.1.3 CORROSION TESTS FROM THE AUTOMOTIVE INDUSTRY

The automotive industry places great demands on its anticorrosion coatings system and has therefore invested a good deal of effort in developing accelerated tests to help predict the performance of paints in harsh conditions. It should be noted that most automotive tests, including the cyclic corrosion tests, have been developed using coatings relevant to automotive application. These are designed to act quite different from protective coatings. Automotive-derived test methods commonly overlook factors critical to protective coatings, such as weathering and UV factors. In addition, automotive coatings have much lower dry-film thickness than do many protective coatings; this is important for mass transport phenomena.

This section is not intended as an overview of automotive industry tests. Some tests that have good correlation to actual field service for cars and trucks, such as the Ford APGE, Nissan CCT-IV, and GM 9540P [6], are not described here. The three tests described here are those believed to be adaptable to heavy maintenance coatings: VDA 621-415, the Volvo Indoor Corrosion Test (VICT), and the Society of Automotive Engineers (SAE) J2334.

15.1.3.1 ISO 11997

For many years, the automotive industry in Germany has used an accelerated test method for organic coatings called the VDA 621-415, which now is an International Organization for Standardization (ISO) standard, ISO 11997; this test has been used as a test for heavy infrastructure paints also. The test consists of 6–12 cycles of neutral salt spray (as per DIN 50021) and 4 cycles in an alternating condensation water climate (as per DIN 50017). The time of wetness of the test is very high, which implies poor correlation to actual service for zinc pigments or galvanized steel. It is expected that zinc will undergo a completely different corrosion mechanism in the nearly constant wetness of the test than the mechanism that takes place in actual field service. The ability of the test to predict the actual performance of zinc-coated substrates and zinc-containing paints must be carefully examined because these materials are commonly used in the corrosion engineering field.

15.1.3.2   Volvo Indoor Corrosion Test or Volvo Cycle

The VICT [7] was developed—despite its name—to simulate the *outdoor* corrosion environment of a typical automobile. Unlike many accelerated corrosion tests, in which the test procedure is developed empirically, the VICT test is the result of a statistical factorial design [8].

In modern automotive painting, all the corrosion protection is provided by the inorganic layers and the thin (ca. 25 μm) electrocoat paint layer. Protection against UV light and mechanical damage is provided by the subsequent paint layers (of which there are usually three). Testing of the *anticorrosion* or electrocoat paint layer can be restricted to a few parameters, such as corrosion-initiating ions (usually chlorides), time of wetness, and temperature. The Volvo test accordingly uses no UV exposure or mechanical stresses; the stresses used are temperature, humidity, and salt solution (sprayed or dipped).

The automotive industry has a huge amount of data for corrosion in various service environments. The VICT has a promising correlation to field data; one criticism that is sometimes brought against this test is that it may tend to produce filiform corrosion at a scribe.

There are four variants of the Volvo cycle, consisting of either constant temperature, together with two levels of humidity, or a constant dew point (i.e., varying temperature and two levels of humidity). The VICT-2 variant, which uses constant temperature and discrete humidity transitions between two humidity levels, is described below.

- *Step I*: Seven hours' exposure at 90% relative humidity (RH) and 35°C constant level
- *Step II*: Continuous and linear change of RH from 90% to 45% RH at 35°C during 1.5 hours
- *Step III*: Two hours' exposure at 45% RH and 35°C constant level
- *Step IV*: Continuous and linear change of RH from 45% to 90% RH at 35°C during 1.5 hours

Twice a week, on Mondays and Fridays, step I is replaced by the following:

- *Step V*: Samples are taken out of the test chamber and submerged in, or sprayed with, 1% (wt) NaCl solution for one hour.
- *Step VI*: Samples are removed from the salt bath; excess liquid is drained off for five minutes. The samples are put back into the test chamber at 90% RH so that they are exposed in wetness for at least seven hours before the drying phase.

Typically, the VICT test is run for 12 weeks. This is a good general test when UV is not expected to be of great importance.

15.1.3.3   SAE J2334

SAE J2334 is the result of a statistically designed experiment using automotive industry substrates and coatings. In the earliest publications about this test, it is also referred to as "PC-4" [6]. The test is based on a 24-hour cycle. Each cycle consists of a 6-hour humidity period at 50°C and 100% RH, followed by a 15-minute salt application, followed by a 17-hour, 45-minute drying stage at 60°C and 50% RH. Typical test duration is 60 cycles; longer cycles have been used for heavier coating weights. The salt concentrations are fairly low, although the solution is relatively complex: 0.5% NaCl + 0.1% $CaCl_2$ + 0.075% $NaHCO_3$.

15.1.4 A TEST TO AVOID: KESTERNICH

In the Kesternich test, samples are exposed to water vapor and sulfur dioxide for 8 hours, followed by 16 hours in which the chamber is open to the ambient environment of the laboratory [2]. This test was designed for bare metals exposed to a polluted industrial environment and is fairly good for this purpose. However, the test's relevance for organically coated metals is highly questionable. Berke and Townsend have also documented poor correlation to field exposure [9]. For the same reason, the similar test ASTM B605 is not recommended for painted steel.

15.1.5 SALT SPRAY TEST

The salt spray (fog) test ASTM B117 ("Standard Practice for Operating Salt Spray [Fog] Testing Apparatus") is one of the oldest corrosion tests still in use. Despite a widespread belief among experts that the salt spray test is of no value in predicting performance, or even relative ranking, of coatings in most applications, it is the most frequently specified test for evaluating paints and substrates.

The salt spray test has such a poor reputation among workers in the field that the word *infamous* is sometimes used as a prefix to the test number. In fact, nearly every peer-reviewed paper published these days on the subject of accelerated testing starts with a condemnation of the salt spray test [10–13]. For example,

- "In fact, it has been recognized for many years that when ranking the performance levels of organic coating systems, there is little if any correlation between results from standard salt spray tests and practical experience" [3].
- "The well-known ASTM B117 salt spray test provides a comparison of cold-rolled and electrogalvanized steel within several hundred hours. Unfortunately, the salt spray test is unable to predict the well-known superior corrosion resistance of galvanized relative to uncoated cold rolled steel sheet" [6].
- "Salt spray provides rapid degradation but has shown poor correlation with outdoor exposures; it often produces degradation by mechanisms different from those seen outdoors and has relatively poor precision" [14].

Many studies comparing salt spray results and actual field exposure have been performed. Coating types, substrates, locations, and length of time have been varied. No correlations have been found to exist between the salt spray and the following service environments:

- Galveston Island, Texas (16 months), 800 m from the sea [15]
- Sea Isle City, New Jersey (28 months), a marine exposure site [16]
- Daytona Beach, Florida (3 years) [17]
- Pulp mills at Lessebo and Skutskar, Sweden, painted hot-rolled steel substrates (four years) [18] and painted aluminum, galvanized steel, and carbon steel substrates (five years) [19]
- Kure Beach, North Carolina, a marine exposure site [20–22]

Appleman and Campbell [23] have examined each of the accelerating stresses in the salt spray test and its effect on the corrosion mechanism compared with outdoor or "real-life" exposure. They found the following flaws in the salt spray test:

1. Constant humid surface
   - Neither the paint nor the substrate experience wet–dry cycles. Corrosion mechanisms may not match those seen in the field; for example, in zinc-rich coatings or galvanized substrates, the zinc is not likely to form a passive film as it does in the field.
   - Water uptake and hydrolysis are greater than in the field.
   - A constant water film with high conductivity is present, which does not happen in the field.
2. Elevated temperature
   - Water, oxygen, and ion transport are greater than in the field.
   - For some paints, the elevated temperature of the test comes close to the glass transition temperature of the binder.
3. *High chloride concentration* (effect on corrosion depends on the type of protection the coating offers)
   - For sacrificial coatings, such as zinc-rich primers, the high chloride content, together with the constant high humidity, means that the zinc is not likely to form a passive film as it does in the field.
   - For inhibitive coatings, chlorides adsorb on the metal surface, where they prevent passivation.
   - For barrier coatings, the osmotic forces are much less than in the field; in fact, they may be reversed completely from that which is seen in reality. In the salt spray test, corrosion at a scribe or defect is exaggeratedly aggressive compared with a scribe under intact paint.

Lyon et al. [24] point out that the high sodium chloride content of the salt spray test can result in corrosion morphologies and behaviors that are not representative of natural conditions. Harrison and Tickle have pointed out that the test is inappropriate for use on zinc—galvanized substrates or primers with zinc phosphate pigments, for example—because, in the constant wetness of the salt spray test, zinc undergoes a corrosion mechanism that it would not undergo in real service [25]. This is a well-known and well-documented phenomenon and is discussed in depth in Chapter 13.

15.1.6 IMPORTANCE OF WET–DRY CYCLING

It has been known for a long time that wet–dry cycling is essential for correlation between accelerated aging tests and field exposure with respect to the corrosion creep result, on both steel and zinc-coated steel [3,24,26–29]. We now know why. The constant wetting in the test differs from the dry–wet cycling the coating will experience in the field. When evaluating the samples from the test, cathodic disbonding (CD) is found in the degradation front, while anodic undermining is found in the field [14]. This has now actually been measured with the scanning Kelvin probe (SKP). Constant wetting results in degradation by CD. Drying is necessary to inverse the potential between the coating damage and the degradation front, to switch from CD to anodic undermining [30]. See Chapter 12 for further discussion of this.

15.1.7 WEATHERING

In UV weathering tests, condensation is alternated with UV exposure to study the effect of UV light on organic coatings. The temperature, amount of UV radiation, length (time) of UV radiation, and length (time) of condensation in the chamber are programmable. Examples of UV weathering tests include QUV-A, QUV-B (®Q-LAB) and xenon tests. Recommended practices for UV weathering are described in the very useful standard ASTM G154.

15.1.8 CONDENSATION OR HUMIDITY

Many tests are based on constant condensation or humidity. Incidentally, constant condensation is not the same as humidity testing. Condensation rates are higher in the former than the latter because, in constant condensation chambers, the back sides of the panels are at room temperature and the painted side faces water vapor at 40°C. This slight temperature differential leads to higher water condensation on the panel. If no such temperature differential exists, the conditions provide humidity testing in what is known as a "tropical chamber." The Cleveland chamber is one example of condensation testing; a salt spray chamber with the salt fog turned off, the heater turned on, and water in the bottom (to generate vapor) is a humidity test.

Constant condensation or humidity testing can be useful as a test for barrier properties of coatings on less than ideal substrates—for example, rusted steel. Any hygroscopic contaminants, such as salts entrapped in the rust, attract water. On new construction, or in the repainting of old construction, where it is possible to blast the steel to Sa2½, these contaminants are not be found. However, for many applications, dry abrasive or wet blasting is not possible, and only handheld tools, such as wire brushes, can be used. These tools remove loose rust but leave tightly adhering rust in place. And, because corrosion-causing ions, such as chloride (Cl⁻), are always at the bottom of corrosion pits, the matrix of tightly adhering rust necessarily contains these hygroscopic contaminants. In such cases, the coating must prevent water from reaching the intact steel. The speed with which blisters develop under the coating in condensation conditions can be an indication of the coating's ability to provide a water barrier and thus protect the steel.

Various standard test methods using constant condensation or humidity testing include ISO 6270 and ISO 11503, the British BS 3900, the North American ASTM D2247 and ASTM D4585, and the German DIN 50017.

15.2 EVALUATION AFTER ACCELERATED AGING

After the accelerated aging, samples should be evaluated for changes. By comparing samples before and after aging, one can find

- Direct evidence of corrosion
- Signs of coating degradation
- Implicit signs of corrosion or failure

The coating scientist uses a combination of techniques for detecting macroscopic and submicroscopic changes in the coating–substrate system. The quantitative and qualitative data this provides must then be interpreted so that a prediction can be made as to whether the coating will fail and, if possible, why.

Macroscopic changes can be divided into two types:

1. Changes that can be seen by the unaided eye or with optical (light) microscopes, such as rust-through and creep from the scribe
2. Large-scale changes that are found by measuring mechanical properties, of which the most important are adhesion to the substrate and the ability to prevent water transport

Changes in both the adhesion values obtained in before and after testing and the failure loci can reveal quite a bit about aging and failure mechanisms. Changes in barrier properties, measured by electrochemical impedance spectroscopy (EIS), are important because the ability to hinder transport of electrolyte in solution is one of the more important corrosion protection mechanisms of the coating.

One may be tempted to include such parameters as loss of gloss or color change as macroscopic changes. However, although these are reliable indicators of UV damage, they are not necessarily indicative of any weakening of the corrosion protection ability of the coating system as a whole, because only the appearance of the topcoat is examined.

Submicroscopic changes cannot be seen with the naked eye or a normal laboratory light microscope, but must instead be measured with advanced electrochemical or spectroscopic techniques. Examples include changes in chemical structure of the paint surface that can be found using Fourier transform infrared (FTIR) spectroscopy or changes in the morphology of the paint surface that can be found using atomic force microscopy (AFM). These changes can yield information about the coating–metal system, which is then used to predict failure, even if no macroscopic changes have yet taken place.

More sophisticated studies of the effects of aging factors on the coating include

- Electrochemical monitoring techniques: alternation current (AC) impedance (EIS), Kelvin probe

- Changes in chemical structure of the paint surface using FTIR or x-ray photoelectron spectroscopy (XPS)
- Morphology of the paint surface using scanning electron microscopy (SEM) or AFM

15.2.1 GENERAL CORROSION

Direct evidence of corrosion can be obtained by macroscopic measurement of creep from the scribe, rust intensity, blistering, cracking, and flaking.

15.2.1.1 Creep from the Scribe

If a coating is properly applied to a well-prepared surface and allowed to cure, then general corrosion across the intact paint surface is not usually a major concern. However, once the coating is scratched and metal is exposed, the situation is dramatically different. An electrochemical cell is established between the exposed steel and the coated metal around, causing corrosion to spread under the coating (Chapter 12). The coating's ability to resist this spread of corrosion is a major concern.

Corrosion that begins in a scratch and spreads under the paint is called *creep* or *undercutting*. Creep is surprisingly difficult to quantify, because it is seldom uniform. Several methods are acceptable for measuring it, for example,

- Maximum one-way creep (probably the most common method), which is used in several standards, such as ASTM S1654
- Summation of creep at 10 evenly spaced sites along the scribe
- Average two-way creep

None of these methods are satisfactory for describing filiform corrosion. The maximum one-way creep and the average two-way creep methods allow measurement of two values: general creep and filiform creep.

15.2.1.2 Other General Corrosion

Blistering, rust intensity, cracking, and flaking are judged in accordance with the standard ISO 4628 or the comparable standard ASTM D610. In these methods, the samples to be evaluated are compared with a set of standard photographs showing various degrees of each type of failure.

For face blistering, the pictures in the ISO standard represent blister densities from 2 to 5, with 5 being the highest density. Blister size is also numbered from 2 to 5, with 5 indicating the largest blister. Results are reported as blister density, followed in parentheses by blister size (e.g., 4(S2) means blister density = 4 and blister size = 2); this is a way to quantify the result, "many small blisters."

For degree of rusting, the response of interest is rust under the paint, or rust bleed-through. Areas of the paint that are merely discolored on the surface by rusty run-off are not counted if the paint underneath is intact. The scale used by ISO 4628 in assigning degrees of rusting is shown in Table 15.1 [31].

Although the ASTM and ISO standards are comparable in methodology, their grading scales run in opposite directions. In measuring rust intensity or blistering,

TABLE 15.1

Degrees of Rusting

Degree	Area Rusted (%)
Ri 0	0
Ri 1	0.05
Ri 2	0.5
Ri 3	1
Ri 4	8
Ri 5	40–50

Source: ISO 4628-3, Designation of degree of rusting, International Organization for Standardization, Geneva, 2016.

the ASTM standard uses 10 for defect-free paint and 0 for complete failure. The ISO standard uses 0 for no defects and the highest score for complete failure.

These standards have faced some criticisms, mainly the following:

- They are too subjective.
- They assume an even pattern of corrosion over the surface.

Proposals have been made to counter the subjective nature of the tests by, for example, adding grids to the test area and counting each square that has a defect. The assumption of an even pattern of corrosion is questioned on the grounds that corrosion, although severe, can be limited to one region of the sample. Systems have been proposed to more accurately reflect these situations, for example, reporting the percentage of the surface that has corrosion and then grading the corrosion level within the affected (corroded) areas. For more information on this, the reader is directed to Appleman's review [2].

15.2.2 ADHESION

Many methods are used to measure adhesion of a coating to a substrate. The most commonly used methods belong to one of the following two groups: direct pull-off (DPO) methods (e.g., ISO 4624) or cross-cut methods (e.g., ISO 2409). The test method must be specified; details of pull-stub geometry and adhesive used in DPO methods are important for comparing results and must be reported.

15.2.2.1 Difficulty of Measuring Adhesion

It is impossible to mechanically separate two well-adhering bodies without deforming them; the fracture energy used to separate them is therefore a function of both the interfacial processes and bulk processes within the materials [32]. In polymers, these bulk processes are commonly a complex blend of plastic and elastic deformation modes and can vary greatly across the interface. This leads to an interesting conundrum: the fundamental understanding of the wetting of a substrate by a liquid coating, and the subsequent adhesion of the cured coating to the substrate, is one of the best-developed areas of coating science, yet methods for the practical measurement of adhesion are comparatively crude and unsophisticated.

It has been shown that experimentally measured adhesion strengths consist of basic adhesion plus contributions from extraneous sources. Basic adhesion is the adhesion that results from the sum of forces between the coating and the substrate; extraneous contributions include internal stresses in the coating and defects or extraneous processes introduced in the coating as a result of the measurement technique itself [32]. To complicate matters, the latter can decrease basic adhesion by introducing new, unmeasured stresses or can increase the basic adhesion by relieving preexisting internal stresses.

The most commonly used methods of detaching coatings are applying a normal force at the interface plane or applying lateral stresses.

15.2.2.2 Direct Pull-Off Methods

DPO methods measure the force per unit area necessary to detach two materials, or the work done (or energy expended) in doing so. DPO methods employ normal forces at the coating–substrate interface plane. The basic principle is to attach a pulling device (a stub or dolly) to the coating by glue, and then to apply a force to it in a direction perpendicular to the painted surface, until either the paint pulls off the substrate or failure occurs within the paint layers (Figure 15.1).

An intrinsic disadvantage of DPO methods is that failure occurs at the weakest part of the coating system. This can occur cohesively within a coating layer; adhesively between coating layers, especially if the glue has created a weak boundary layer within the coating; or adhesively between the primer layer and the metal substrate, depending on which is the weakest link in the system. Therefore, adhesion of the primer to the metal is not necessarily what this method measures, unless it is at this interface that the adhesion is the weakest.

DPO methods suffer from some additional disadvantages:

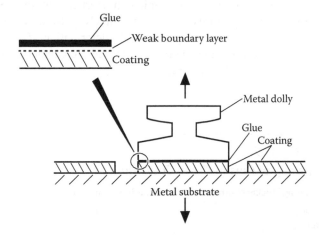

FIGURE 15.1   DPO adhesion measurement.

- Tensile tests usually involve a complex mixture of tensile and shear forces just before the break, making interpretation difficult.
- Stresses produced in the paint layer during setting of the adhesive may affect the values measured (a glue–paint interaction problem).
- Nonuniform tensile load distributions over the contact area during the pulling process may occur. Stress concentrated in a portion of the contact area leads to failure at these points at lower values than would be seen under even distribution of the load. This problem usually arises from the design of the pulling head.

Unlike lateral stress methods, DPO methods can be used on hard or soft coatings. As previously mentioned, however, for a well-adhering paint, these methods tend to measure the cohesive strength of the coating, rather than its adhesion to the substrate.

With DPO methods, examination of the ruptured surface is possible, not only for the substrate but also for the test dolly. A point-by-point comparison of substrate and dolly surfaces makes it possible to fairly accurately determine interfacial and cohesive failure modes.

15.2.2.3 Lateral Stress Methods

Methods employing lateral stresses to detach a coating include bend or impact tests and scribing the coating with a knife, as in the cross-cut test.

In the cross-cut test, which is the most commonly used of the lateral stress methods, knife blades scribe the coating down to the metal in a grid pattern. The spacing of the cuts is usually determined by the coating thickness. Standard guidelines are given in Table 15.2. The amount of paint removed from the areas adjacent to, but not touched by, the blades is taken as a measurement of adhesion. A standard scale for evaluation of the amount of flaking is shown in Table 15.3.

Analysis of the forces involved is complex because both shear and peel can occur in the coating. The amount of shearing and peeling forces created at the knife tip depends not only on the energy with which the cuts are made (i.e., force and speed of scribing) but also on the mechanical properties—plastic versus elastic deformation—of the coating. For example, immediately in front of the knife edge, the upper surface of the paint undergoes plastic deformation. This deformation produces a shear force down at the coating–metal interface, underneath the rim of indentation in front of the knife edge [32].

TABLE 15.2
Spacing of Cuts in Cross-Cut Adhesion

Coating Thickness (μm)	Spacing of the Cuts (mm)
<60	1
60–120	2
>120	3

TABLE 15.3

Evaluation of the Amount of Flaking

Grade	Description
0	The cuts are very sharp. No material has flaked.
1	Somewhat uneven cuts. Detachment of small flakes of the coating at the intersections of the cuts.
2	Clearly uneven cuts. The coating has flaked along the edges or at the intersections of the cuts.
3	Very uneven cuts. The coating has flaked along the edges of the cuts partly or wholly in large ribbons, or it has flaked partly or wholly on different parts of the squares. A cross-cut area of maximum 35% may be affected.
4	Severe flaking of material. The coating has flaked along the edges of the cuts in large ribbons, or some squares have been detached partly or wholly. A cross-cut area of maximun 65% may be affected.
5	A cross-cut area greater than 65% is affected.

A major drawback to methods using lateral stresses is that they are extremely dependent on the mechanical properties of the coating, especially how much plastic and elastic deformation the coating undergoes. Paul has noted that many of these tests result in cohesive cracking of coatings [32]. Coatings with mostly elastic deformation commonly develop systems of cracks parallel to the metal–coating interface, leading to flaking at the scribe and poor test results. Coatings with a high proportion of plastic deformation, on the other hand, perform well in this test—even though they may have much poorer adhesion to the metal substrate than do hard coatings.

Elastic deformation means that little or no rounding of the material occurs at the crack tip during scribing. Almost all the energy goes into crack propagation. As the knife blade moves, more cracks in the coating are initiated further down the scribe. These propagate until two or more cracks meet and lead to flaking along the scribe. The test results can be misleading; epoxies, for example, usually perform worse than softer alkyds in cross-cut testing, even though, in general, they have much stronger adhesion to metal.

For very hard coatings, scribing down to the metal may not be possible. Use of the cross-cut test appears to be limited to comparatively soft coatings. Because the test is very dependent on deformation properties of the coatings, comparing cross-cut results of different coatings to each other is of questionable value. However, the test may have some value in comparing the adhesion of a single coating to various substrates or pretreatments.

15.2.2.4 Important Aspects of Adhesion

The failure loci—where the failure occurred—can yield very important information about coating weaknesses and eventual failures. Changes in failure loci related to aging of a sample are especially revealing about what is taking place within and under a coating system.

Adhesion measurements are performed to gain information regarding the mechanical strengths of the coating–substrate bonds and the deterioration of these bonds when the coatings undergo environmental stresses. A great deal of work has been done to develop better methods for measuring the strengths of the initial coating–substrate bonds.

By comparison, little attention has been given to using adhesion tests to obtain information about the mechanism of deterioration of either the coating or its adhesion to the metal. This area deserves greater attention because studying the failure loci in adhesion tests before and after weathering can yield a great deal of information about why coatings fail.

Finally, it is important to remember that adhesion is only one aspect of corrosion protection. At least one study shows that the coating with the best adhesion to the metal did not provide the best corrosion protection [33]. Also, studies have found that there is no obvious relationship between initial adhesion and wet adhesion [34].

15.2.3 INTERNAL STRESS IN PAINT FILMS

ASTM D6991 describes measuring internal stress in organic coatings with a cantilever method, based on work by Perera and van Eynde [35] and Korobov and Salem [36]. The coating is applied to one side of a thin metal strip. The strip can be of, for example, stainless steel or aluminum and should be less than 300 μm thick, 10–20 mm wide, and 100–200 mm long. Internal stress in the film will make the strip curl, and the deflection is used to calculate the internal stress according to Equation 15.1 [37]:

$$S = \frac{h E_s t^3}{3L^2 c (t+c)(1-\gamma_s)} \tag{15.1}$$

where S is internal stress in megapascals; h is deflection of the strip in millimeters; E_s is the modulus of elasticity of the strip in megapascals; γ_s is the Poisson ratio of the strip; and L is the length of the strip, t is the thickness of the strip, and c is the coating thickness, all in millimeters. This equation is only valid when the coating is significantly thinner than the strip, however. If thicker coatings are tested, Equation 15.2 must be used [37]:

$$S = \frac{h E_s t^3}{3L^2 c (t+c)(1-\gamma_s)} + \frac{h E_c (t+c)}{L^2 (1-\gamma_c)} \tag{15.2}$$

Here, E_c is modulus of elasticity of the coating and γ_c is the Poisson ratio for the coating. These values are usually not known, which complicates the method somewhat. The Poisson ratio of the coating can be assumed to be similar to that of the substrate, but the coating modulus of elasticity can vary over a wide range and must be measured.

15.3   ACCELERATED TESTING OF COATINGS
FOR IMMERSION SERVICE

As for aging testing, there is a large number of test methods for immersed coatings, that is, CD tests. Camron et al. have given an overview [38]. Many of the tests have specifically been developed for testing coatings on buried or immersed pipelines, but technically there is no difference between CD on pipelines and structural steel. The various tests differ in electrolyte composition, temperature, applied potential, and duration. Only three methods will be mentioned here, an ambient temperature test, a flexible temperature test, and a high-temperature test, to illustrate some of the issues with CD testing.

15.3.1   ISO 15711

This test was developed for structural steel exposed in seawater at ambient temperature and is referred to by ISO 20340 (and thereby NORSOK M-501 also) for CD testing. The test is performed in natural or artificial seawater at ambient laboratory temperature and a cathodic potential of −1.05 V versus Ag/AgCl. This means that the test conditions are very similar to field conditions, and the test is hardly accelerated at all. Very good correlation to field performance should therefore be expected. The low level of acceleration is also the main drawback with the test—a long duration is required in order to distinguish between coatings. The specified test duration is a minimum of 26 weeks.

15.3.2   NACE TM0115

This is the most recently issued CD test, published by NACE in 2015. The purpose for developing yet another CD test was to have a more accurately specified procedure to reduce variation between test labs. For example, how to maintain constant oxygen concentration in the electrolyte, positioning of the reference electrode, temperature measurement, insulation of the counterelectrode, how to prepare the coating holiday, cathodic current monitoring, and how to peel off the disbonded coating at the end of the test are specified in this standard, while neglected or only partly described in other standards. At the same time, the test is also flexible with respect to parameters that will not affect the result, for example, how to heat the sample when testing at elevated temperature and whether to perform the test with an attached cell on the sample surface or samples immersed in the test electrolyte.

The test is performed in 3% NaCl with a moderately accelerated potential of 1.38 V versus Ag/AgCl (−1.50 V vs. Cu/CuSO$_4$). The duration is 30 days. The test temperature is selected according to the field conditions in which the coating will be exposed. For testing coatings for onshore pipelines up to 95°C, sample and electrolyte are kept at the same temperature. When testing above 95°C, only the steel is heated to the test temperature, while the electrolyte is kept at 95°C. When testing coatings for offshore pipelines at elevated temperature, the electrolyte temperature may be kept at 30°C, since seawater convection will have a cooling effect offshore.

15.3.3   NORSOK M-501 HIGH-TEMPERATURE CD TEST

This test was developed for evaluation of coatings for submerged high-temperature structural steel surfaces. The test is performed with artificial seawater at a moderately accelerated potential of 1.2 V versus Ag/AgCl. The steel is heated to the specified field temperature, while the electrolyte is kept at 30°C. At steel temperatures above 100°C, the test electrolyte is pressurized in order to prevent evaporation of the coating in the crevice under the disbonded coating.

15.4   ADVANCED METHODS FOR INVESTIGATION OF PROTECTIVE PROPERTIES AND DEGRADATION MECHANISMS

15.4.1   BARRIER PROPERTIES

Coatings, being polymer based, are naturally highly resistant to the flow of electricity. This fact is utilized to measure water uptake by and transport through the coating. The coating itself does not conduct electricity; any current passing through it is carried by electrolytes in the coating. Measuring the electrical properties of the coating makes it possible to calculate the amount of water present (called *water content* or *solubility*) and how quickly it moves (called *diffusion coefficient*). The technique used to do this is EIS.

An intact coating is described in EIS as a general equivalent electrical circuit, also known as the *Randles model* (Figure 15.2). As the coatings become more porous or local defects occur, the model becomes more complex (Figure 15.3). Circuit A in Figure 15.3 is the more commonly used model; it is sometimes referred to as the *extended Randles model* [39,40].

EIS is an extremely useful technique in evaluating the ability of a coating to protect the underlying metal, and an ISO standard has been issued [41]. It is frequently used as a before-and-after test because it is used to compare the water content and diffusion coefficient of the coating before and after aging (accelerated or natural exposure). Królikowska [42] has suggested that for a coating to provide corrosion protection to steel, it should have an initial impedance of at least $10^8/cm^2$, a

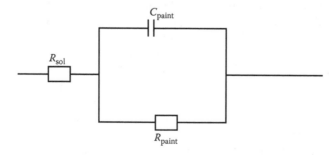

FIGURE 15.2 Equivalent electric circuit to describe an intact coating. R_{sol} is the solution resistance, C_{paint} is the capacitance of the paint layer, and R_{paint} is the resistance of the paint layer.

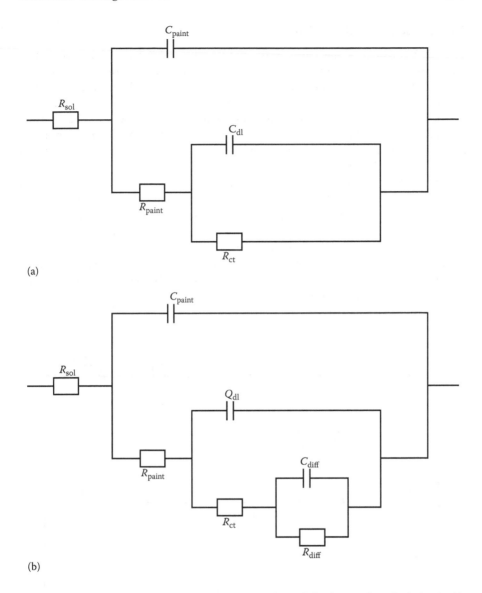

FIGURE 15.3 Equivalent electric circuits to describe a defective coating. C_{dl} is the double layer capacitance, R_{ct} is the charge transfer resistance of the corrosion process, Q_{dl} is the constant phase element, C_{diff} is the diffuse layer capacitance, and R_{diff} is the diffuse layer resistance.

value also suggested by others [39], and that after aging, the impedance should have decreased by no more than three orders of magnitude. Sekine has reported blistering when the coating resistance falls below $10^6/cm^2$, regardless of coating thickness [43].

For more in-depth reviews of the fundamental concepts and models used in EIS to predict coating performance, the reader is directed to the research of Kendig and Scully [44] and Walter [45–47].

15.4.2   SCANNING KELVIN PROBE

The SKP provides a measure of the Volta potential (work function) that is related to the corrosion potential of the metal, without touching the corroding surface [48]. The technique can give a corrosion potential distribution below highly isolating polymer films. The lateral resolution depends on the distance between the probe and the steel substrate, that is, the thickness of the coating. The SKP is an excellent research tool to study the initiation of corrosion at the metal–polymer interface, as demonstrated in several of the investigations referred to in Chapters 12 and 13.

Figures 15.4 and 15.5 show the Volta potential distribution for a coil-coated sample before and after five weeks of weakly accelerated field testing [49]. In the "after" figure, a large zone at low potential (–850 to –750 mV/NHE) can be clearly seen. Delamination, corrosion, or both are occurring at the transition area between the "intact" metal–polymer interface (zones at higher potential values, –350 to –200 mV/NHE) and more negative electrode potentials. The corrosion that is starting here after five weeks will not be visible as blisters for nearly two years at the Bohus-Malmön coastal station in Sweden [49].

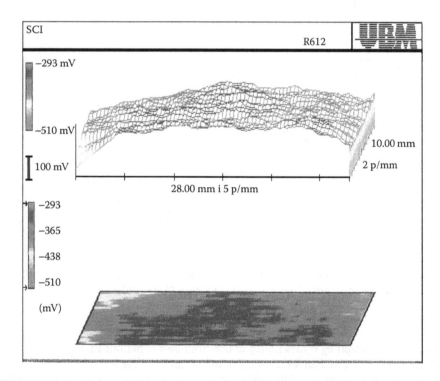

FIGURE 15.4   Volta distribution (mV) of coated steel before exposure. (From Forsgren, A., and Thierry, D., Corrosion properties of coil-coated galvanized steel, using field exposure and advanced electrochemical techniques, Report 2001:4E, Swedish Corrosion Institute, Stockholm, 2001. Photo courtesy of Swedish Corrosion Institute.)

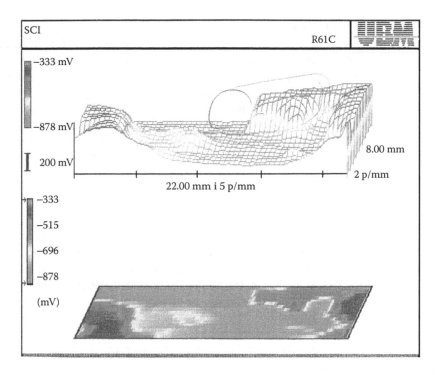

FIGURE 15.5 Volta distribution (mV) of coated steel before (top) and after (bottom) five weeks' weakly accelerated field exposure. (From Forsgren, A., and Thierry, D., Corrosion properties of coil-coated galvanized steel, using field exposure and advanced electrochemical techniques, Report 2001:4E, Swedish Corrosion Institute, Stockholm, 2001. Photo courtesy of Swedish Corrosion Institute.)

15.4.3 SCANNING VIBRATING ELECTRODE TECHNIQUE

The scanning vibrating electrode technique (SVET) is used to quantify and map localized corrosion. The instrument moves a vibrating probe just above (100 µm or less) the sample surface, measuring and mapping the electric fields that are generated in the adjacent electrolyte as a result of localized electrochemical or corrosion activity. It is a well-established tool in researching localized events, such as pitting corrosion, intergranular corrosion, and coating defects. The SVET, which gives a two-dimensional distribution of current, is similar in many respects to the SKP; in fact, some instrument manufacturers offer a combined SVET-SKP system.

15.4.4 ADVANCED ANALYTICAL TECHNIQUES

For the research scientist or the well-equipped failure analysis laboratory, several advanced analytical techniques can prove useful in studying protective coatings. Many such techniques are based on detecting charged particles that come from, or interact with, the surface in question. These require high (10^{-5} or 10^{-7} torr) or ultrahigh vacuum (less than 10^{-8} torr), which means that samples cannot be studied *in situ* [50].

15.4.4.1 Scanning Electron Microscopy

Unlike optical microscopes, SEM does not use light to examine a surface. Instead, SEM sends a beam of electrons over the surface to be studied. These electrons interact with the sample to produce various signals: x-rays, back-scattered electrons, secondary electron emissions, and cathode luminescence. Each of these signals has slightly different characteristics when they are detected and photographed. SEM has very high depth of focus, which makes it a powerful tool for studying the contours of surfaces.

Electron microscopes used to be found only in research institutes and more sophisticated industrial laboratories. They have now become more ubiquitous; in fact, they are an indispensable tool in advanced failure analysis and are found in most any laboratory dealing with material sciences.

15.4.4.2 Atomic Force Microscopy

AFM provides information about the morphology of a surface. Three-dimensional maps of the surface are generated, and some information of the relative hardness of areas on the surface can be obtained. AFM has several variants for different sample surfaces, including contact mode, tapping mode, and phase contrast AFM. Soft polymer surfaces, such as those found in many coatings, tend to utilize tapping mode AFM.

In waterborne paint research, AFM has proven an excellent tool for studying coalescence of latex coatings [50–53]. It has also been used to study the initial effect of waterborne coatings on steel before film formation can occur, as shown in Figures 15.6 and 15.7 [54].

15.4.4.3 Infrared Spectroscopy

Infrared spectroscopy is a family of techniques that can be used to identify chemical bonds. When improved by Fourier transform mathematical techniques, the

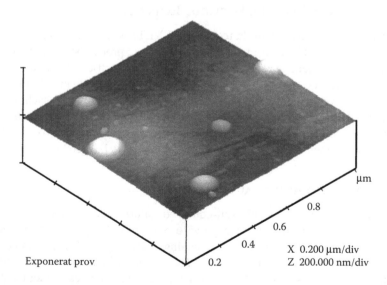

FIGURE 15.6 Example of AFM imaging. (Photo courtesy of Swedish Corrosion Institute.)

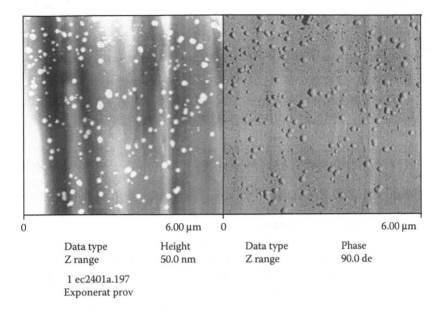

Data type	Height	Data type	Phase
Z range	50.0 nm	Z range	90.0 de

1 ec2401a.197
Exponerat prov

FIGURE 15.7   Example of AFM imaging. (Photo courtesy of Swedish Corrosion Institute.)

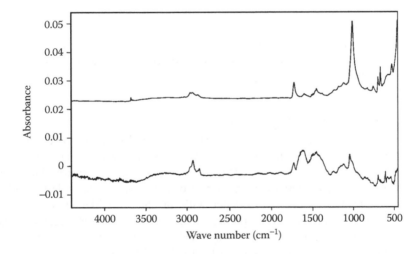

FIGURE 15.8   Example of FTIR fingerprinting. (Photo courtesy of Swedish Corrosion Institute.)

resulting test is known as FTIR. An FTIR scan can be used to identify compounds in the same way as fingerprints are used to identify humans: an FTIR scan of the sample is compared with the FTIR scans of "known" compounds. If a positive match is found, the sample has been identified; an example is shown in Figure 15.8. Not surprisingly, FTIR results are sometimes called "fingerprints" by analytical chemists.

The most important FTIR techniques include

- Attenuated total reflectance (ATR), in which a sample is placed in close contact with the ATR crystal. ATR is excellent on smooth surfaces that do not degrade during the test.
- Diffuse internal reflectance (DRIFT). DRIFT uses potassium bromide pellets for sample preparation and therefore has certain limitations in use with hygroscopic materials.
- Photoacoustic spectroscopy (PAS). In PAS, the sample surface absorbs radiation, heats up, and gives rise to thermal waves. These cause pressure variations in the surrounding gas, which are transmitted to a microphone—hence the acoustic signal [55].

15.4.4.4 Electron Spectroscopy

Electron spectroscopy is a type of chemical analysis in which a surface is bombarded with particles or irradiated with photons so that electrons are emitted from it. Broadly speaking, different elements emit electrons in slightly different ways; so an analysis of the patterns of electrons emitted—in particular, the kinetic energy of the electron in the spectrometer and the energy required to knock it off the atom (binding energy)—can help identify the atoms present in the sample.

There are several types of electron spectroscopy techniques, each differing in their irradiation sources. The one most important to coatings research, XPS (or electron spectroscopy for chemical analysis [ESCA]), uses monochromatic x-rays. XPS can identify elements (except hydrogen and helium) located in the top 1–5 nm of a surface. It also gives information about oxidation states. It is a powerful research tool that has, for example, been used to characterize surfaces after adhesive fracture [56,57].

15.4.4.5 Electrochemical Noise Measurement

Electrochemical noise measurement (ENM) has attracted attention since it was first applied to anticorrosion coatings in the late 1980s [58]. The noise consists of fluctuations in the current or potential that occur during the course of corrosion. The underlying idea is that these fluctuations in current or potential are not entirely random. An unavoidable minimum noise associated with current flow will always be random. However, if this minimum can be predicted for an electrochemical reaction, then analysis of the remainder of the noise may yield information about other processes, such as pitting corrosion, mass transport fluctuations, and the formation of bubbles (i.e., hydrogen formed at the cathode).

The theoretical treatment of electrochemical noise is not complete. There does not yet seem to be consensus on which signal analysis techniques are most useful. It is fairly clear, however, that understanding of ENM requires a good working knowledge of statistics; anyone setting out to master the technique must steel themselves to hear of kurtosis, skewness, and block averages rather frequently.

In the future, this technique may become a standard research tool for localized corrosion processes that give strong electrochemical noise signals, such as microbial corrosion and pitting corrosion.

15.5 CALCULATING THE AMOUNT OF ACCELERATION AND CORRELATIONS

Accelerated tests are most commonly used in one of two ways:

1. To compare or rank a series of samples in order to screen out unsuitable coatings or substrates (or conversely, in order to find the most applicable ones)
2. To predict whether a coating–substrate combination will give satisfactory performance in the field—and for how long

This requires that it be possible to calculate both the amount of acceleration the test causes and how uniform this amount of acceleration is over a range of substrates and coatings.

In order to be useful in comparing different coating systems or substrates, an accelerated test must cause even acceleration of the corrosion process among all the samples being tested. Different paint types have different corrosion protection mechanisms; therefore, accentuating one or more stresses—such as heat or wet time—can be expected to produce different amounts of acceleration of corrosion among a group of coatings. The same holds true for substrates. As the stress or stresses are further accentuated—higher temperatures, more wet time, more salt, and more UV light—the variation in the corrosion rates for different coatings or different substrates increases. Three samples sitting side by side in an accelerated test, for example, may have 3×, 2×, and 8× acceleration rates due to different vulnerabilities in different coatings. The problem is that the person performing the test, of course, does not know the acceleration rate for each sample. This can lead to incorrect ranking of coatings or substrates when the accelerated test is completed.

The problem for any acceleration method, therefore, is to balance the amount of acceleration obtained with the variation (among different coatings or substrates). The variation should be minimal and the acceleration should be maximal; this is not trivial to evaluate because, in general, a higher acceleration can be expected to produce more variation in the acceleration rate for the group of samples.

15.5.1 ACCELERATION RATES

The amount of acceleration provided by a laboratory test could be considered quite simply the ratio of the amount of corrosion seen in the laboratory test to the amount seen in field exposure (also known as "reference") over a comparable time span. It is usually reported as $2X$, $10X$, and so on, where $2X$ would be corrosion in the lab occurring twice as quickly as in the field, as shown in Equation 15.3:

$$A = \frac{X_{\text{accel}}}{X_{\text{field}}} \cdot \frac{t_{\text{field}}}{t_{\text{accel}}} \qquad (15.3)$$

where:

   A is the rate of acceleration

   X_{accel} is the response (creep from the scribe) from the accelerated test

   X_{field} is the response from field exposure

   t_{accel} is the duration of the acceleration test

   t_{field} is the duration of the field exposure

For example, after running test XYZ in the lab for five weeks, 4 mm creep from the scribe was seen on a certain sample. After two years' outdoor exposure, an identical sample showed 15 mm creep from the scribe. The rate of acceleration, A, could be calculated as

$$A = \frac{(4 \text{ mm}/5 \text{ weeks})}{(15 \text{ mm}/104 \text{ weeks})} = 5.5 \tag{15.4}$$

15.5.2 CORRELATION COEFFICIENTS OR LINEAR REGRESSIONS

Correlation coefficients can be considered indicators of the uniformity of acceleration within a group of samples. Correlations by linear least-square regression are calculated for data from samples run in an accelerated test versus the response of identical samples in a field exposure. A high correlation coefficient is taken as an indication that the test accelerates corrosion more or less to the same degree for all samples in the group. One drawback of correlation analyses that use least-square regression is that they are sensitive to the distribution of data [59].

15.5.3 MEAN ACCELERATION RATIOS AND COEFFICIENT OF VARIATION

Another interesting approach to evaluating field data versus accelerated data is the mean acceleration ratio and coefficient of variation [59].

To compare data from a field exposure to data from an accelerated test for a set of panels, the acceleration ratio for each type of material (i.e., coating and substrate) is calculated by dividing the average result from the accelerated test by the corresponding reference value, usually from field exposure. These results are then summed up for all the panels in the set and divided by the number of panels in the set to give the mean acceleration ratio. That is,

$$MVQ = \frac{\sum_{i=1}^{n} \frac{X_{i,accel}}{X_{i,field}}}{n} +/- \sigma_{n-1} \tag{15.5}$$

where:

   MVQ is the mean value of quotients

   $X_{i,accel}$ is the response (creep from the scribe) from the accelerated test for each sample i

$X_{i,\text{field}}$ is the response from field exposure for each sample i
 n is the number of samples in the set [59]

This is used to normalize the standard deviation by dividing it by the mean value (MVQ):

$$\text{Coefficient of variation} = \frac{\sigma_{n-1}}{MVQ}$$

$$\text{Test acceleration} = MVQ \cdot \frac{t_{\text{field}}}{t_{\text{accel}}} \tag{15.6}$$

The coefficient of variation combines the amount of acceleration provided by the test with how uniformly the corrosion is accelerated for a set of samples. It is desirable, of course, for an acceleration test to accelerate the corrosion rate more or less uniformly for all the samples; that is, the standard deviation should be as low as possible. It follows naturally that the ratio of deviation to mean acceleration should be as close to 0 as possible. A high coefficient of variation means that, for each set of data, there is more spread in the amount of acceleration than there is actual acceleration.

REFERENCES

1. Goldie, B. *Prot. Coat. Eur.* 1, 23, 1996.
2. Appleman, B. *J. Coat. Technol.* 62, 57, 1990.
3. Skerry, B.S., et al. *J. Coat. Technol.* 60, 97, 1988.
4. Knudsen, O.Ø., et al. *J. Prot. Coat. Linings* 18, 52, 2001.
5. Knudsen, O.Ø. Review of coating failure incidents on the Norwegian continental shelf since the introduction of NORSOK M-501. Presented at CORROSION/2013. Houston: NACE International, 2013, paper 2500.
6. Townsend, H. Development of an improved laboratory corrosion test by the automotive and steel industries. Presented at the 4th Annual ESD Advanced Coatings Conference. Detroit: Engineering Society of Detroit, 1994.
7. Volvo. Corporate Standard STD 423-0014. Accelerated corrosion test. Gothenburg, Sweden: Volvo Group, 2003.
8. Ström, M. Presented at Proceedings of the Conference on Automotive Corrosion and Prevention. Volvo laboratory study of zinc-coated steel sheet: Corrosion behavior studied by a newly developed multifactor indoor corrosion test, Warrendale, PA: Society of Automotive Engineers, 1989, paper 890705.
9. Berke, N., and H. Townsend. *J. Test. Eval.* 13, 74, 1985.
10. Funke, W. *J. Oil Color Chem. Assoc.* 62, 63, 1979.
11. Lambert, M.R., et al. *Ind. Eng. Chem. Prod. Res. Dev.* 24, 378, 1985.
12. Lyon, S.B., et al. *CORROSION* 43, 719, 1987.
13. Timmins, F.D. *J. Oil Color Chem. Assoc.* 62, 131, 1979.
14. Appleman, B. *J. Prot. Coat. Linings* 6, 71, 1989.
15. Struemph, D.J., and J. Hilko. *IEEE Trans. Power Deliv.* PWRD-2, 823, 1987.
16. Chong, S. L., *J. Prot. Coat. Linings* 14, 20, 1997.
17. Rommal, H.E.G., et al. Accelerated test development for coil-coated steel building panels. Presented at CORROSION/1998. Houston: NACE International, 1998, paper 356.

18. Forsgren, A., and S. Palmgren. Salt spray test vs. field results for coated samples: Part I. Report 1998:4E. Stockholm: Swedish Corrosion Institute, 1998.

19. Forsgren, A., and S. Palmgren. Salt spray test vs. field results for coated samples: Part II. Report 1998:6E. Stockholm: Swedish Corrosion Institute, 1998.

20. Appleman, B.R., et al. Performance of alternate coatings in the environment (PACE). Vol. I: Ten year field data. Report FHWA-RD-89-127. Washington, DC: U.S. Federal Highway Administration, 1989.

21. Appleman, B.R., et al. Performance of alternate coatings in the environment (PACE). Vol. II: Five year field data and bridge data of improved formulations. Report FHWA-RD-89-235. Washington, DC: U.S. Federal Highway Administration, 1989.

22. Appleman, B.R., et al. Performance of alternate coatings in the environment (PACE). Vol. III: Executive summary. Report FHWA-RD-89-236. Washington, DC: U.S. Federal Highway Administration, 1989.

23. Appleman, B.R., and P.G. Campbell. *J. Coat. Technol.* 54, 17, 1982.

24. Lyon, S.B., et al. Materials evaluation using wet-dry mixed salt spray tests. In *New Methods for Corrosion Testing of Aluminum Alloys, ASTM STP 1134*, ed. V.S. Agarwala and G.M. Ugiansky. Philadelphia: American Society for Testing and Materials, 1992.

25. Harrison, J.B., and T.C. Tickle. *J. Oil Color Chem. Assoc.* 45, 571, 1962.

26. Simpson, C.H., et al. *J. Prot. Coat. Linings* 8, 28, 1991.

27. Nowak, E.T., et al. A comparison of corrosion test methods for painted galvanized steel. SAE Technical Paper Series, paper 820427. Warrendale, PA: Society of Automotive Engineers, 1982.

28. Smith, D.M., and G.W. Whelan. Corrosion studies of painted automotive substrates—Research in progress. SAE Technical Paper Series, paper 870646. Warrendale, PA: Society of Automotive Engineers, 1987.

29. Standish, J.V., et al. The corrosion behavior of galvanized and cold rolled steels. SAE Technical Paper Series, paper 831810. Warrendale, PA: Society of Automotive Engineers, 1983.

30. Nazarov, A., and D. Thierry. *CORROSION* 66, 0250041, 2010.

31. ISO 4628-3. Paints and varnishes—Evaluation of degradation of coatings—Designation of quantity and size of defects, and of intensity of uniform changes in appearance—Part 3: Assessment of degree of rusting. Geneva: International Organization for Standardization, 2016.

32. Paul, S. *Surface Coatings Science and Technology*. Chichester: John Wiley & Sons, 1996.

33. Dickie, R.A. *Prog. Org. Coat.* 25, 3, 1994.

34. Walker, P. *Paint Technol.* 31, 22, 1967.

35. Perera, D., and D. van Eynde. *J. Coat. Technol.* 53, 39, 1981.

36. Korobov, Y., and L. Salem. *Mater. Perform.* 29, 30, 1990.

37. Corcoran, E.M. *J. Paint Technol.* 41, 635, 1969.

38. Cameron, K., et al. Critical evaluation of international cathodic disbondment test methods. Presented at CORROSION/2005. Houston: NACE International, 2005, paper 5029.

39. Lavaert, V., et al. *Prog. Org. Coat.* 38, 213, 2000.

40. Özcan, M., et al. *Prog. Org. Coat.* 44, 279, 2002.

41. ISO 16773-2. Electrochemical impedance spectroscopy (EIS) on high impedance coated specimens—Part 2: Collection of data. Geneva: International Organization for Standardization, 2007.

42. Królikowska, A. *Prog. Org. Coat.* 39, 37, 2000.

43. Sekine, I. *Prog. Org. Coat.* 31, 73, 1997.

44. Kendig, M., and J. Scully. *CORROSION* 46, 22, 1990.

45. Walter, G.W. *Corros. Sci.* 32, 1041, 1991.

46. Walter, G.W. *Corros. Sci.* 32, 1085, 1991.
47. Walter, G.W. *Corros. Sci.* 32, 1059, 1991.
48. Stratmann, M., et al. *Corros. Sci.* 30, 715, 1990.
49. Forsgren, A., and D. Thierry. Corrosion properties of coil-coated galvanized steel, using field exposure and advanced electrochemical techniques. Report 2001:4E. Stockholm: Swedish Corrosion Institute, 2001.
50. Gilicinski, A.G., and C.R. Hegedus. *Prog. Org. Coat.* 32, 81, 1997.
51. Gerharz, B., et al. *Prog. Org. Coat.* 32, 75, 1997.
52. Joanicot, M., et al. *Prog. Org. Coat.* 32, 109, 1997.
53. Tzitzinou, A., et al. *Prog. Org. Coat.* 35, 89, 1999.
54. Forsgren, A., and D. Persson. Changes in the surface energy of steel caused by acrylic waterborne paints prior to cure. Report 2000:5E. Stockholm: Swedish Corrosion Institute, 2000.
55. Almeida, E., et al. *Prog. Org. Coat.* 44, 233, 2002.
56. Watts, J.F., and J.E. Castle. *J. Mater. Sci.* 18, 2987, 1983.
57. Watts, J.F., and J.E. Castle. *J. Mater. Sci.* 19, 2259, 1984.
58. Jamali, S.S., and D.J. Mills. *Prog. Org. Coat.* 95, 26, 2016.
59. Ström, M., and G. Ström. A statistically designed study of atmospheric corrosion simulating automotive field conditions under laboratory conditions, paper 932338. SAE Technical Paper Series. Warrendale, PA: Society of Automotive Engineers, 1993.

Index

Milton Keynes UK
Ingram Content Group UK Ltd.
UKHW040109071024
449327UK00019B/934